Learning Physiology through MCQ

Learning Physiology through MCQ

A Comprehensive Text

Mary L. Forsling
Department of Physiology,
University College and Middlesex School of Medicine,
London, UK.

A Wiley Medical Publication

JOHN WILEY & SONS
Chichester · New York · Brisbane · Toronto · Singapore

Library of Congress Cataloging-in-Publication Data:

Forsling, Mary L.
Learning physiology through MCQ.

(A Wiley medical publication)
Includes index.
1. Human physiology. I. Title. II. Series.
QP34.5.F67 1988 612 87-29594

ISBN 0 471 91026 0 (pbk.)

British Library Cataloguing in Publication Data:

Forsling, Mary L.
Learning physiology through MCQ: a comprehensive text.
1. Human physiology—Problems, exercises, etc.
I. Title
612'.0076 QP40

ISBN 0 471 91026 0

Printed and bound in Great Britain

Contents

Preface

In physiology examinations—whether for medical, dental or science students—multiple choice questions (MCQs) often form one component of the test; others probably include essay questions, short answer questions or problem-solving exercises. MCQs, however, may also be useful in self-assessment or as a teaching aid. Working through a series of questions will indicate areas of weakness. MCQs test factual recall and in mastering any subject a certain number of basic facts are required. MCQs can also aid in comprehension. For example, deducing the relationship between two changes hitherto unconsidered by the student, will require them to think more fully about a given system. To aid in this exercise of comprehension and possibly in revision this text provides background information, as well as detailed explanation with each question, rather than just a brief answer.

Each question has been chosen to cover a given topic in physiology and in general the information refers to data obtained in man. Most major aspects of physiology are dealt with, being organized in sequence under the headings general principles of physiology, organ systems and integrated systems, so that it is possible to use consecutive explanatory sections as a basic primer in physiology. However, by the nature of the book, it cannot be entirely comprehensive. The format also means that figures are limited and have only been included where complex relationships need to be described. In preparing the relatively short explanations some simplification has been necessary, but not, it is hoped, at the expense of accuracy. With regard to this latter point, I am grateful to my colleagues at University College for reading many of the sections and making helpful comments. In order that the questions may be chosen at random as well as being attempted in sequence, each explanation is made as independent as possible; inevitably this leads to some duplication. Even though the questions have been written with care and many have been tested, some ambiguities may still occur, but this is less important when the questions are used as teaching exercises as opposed to examination material.

The format of the MCQs is that commonly used in examinations: a stem followed by five items, each of which may be true or false. At the end of each question is the abbreviated answer T or F. In tackling MCQ papers under examination conditions it is important to allocate time with care. The question should be read carefully and one should concentrate on each individual statement (a stem and a single item taken together). The instructions on the question sheet should be read with care and the response correctly transferred to the answer sheet; many errors can occur at this stage! It is wise to leave time to go over the answers again, in particular going back to difficult questions. However, repeated review of the answers is often counterproductive. It is dangerous to calculate the likely score as one goes along in order to gain an estimate of how many responses may be omitted, as one invariably overestimates the score. It is not advisable simply to guess an answer as there is a 50 per cent chance of being wrong. Time should be taken to think out the response, drawing on first principles and reasoning power and delving into memory stores. Each question should be taken at face value. It is not necessary to look for hidden meanings or traps. In approaching an MCQ paper there is no substitute for sound knowledge and it is hoped this text will aid in achieving this.

General Physiology and Body Fluids

MCQ 1

The plasma membrane

1. acts as a passive barrier to most molecules.
2. is freely permeable to most ions.
3. comprises a bilipid layer.
4. contains specific proteins.
5. contains an electric charge.

The cell membrane or plasma membrane is composed of a double layer of phospholipid molecules with the polar or hydrophilic region lying at the surface and the non-polar fatty acids being oriented towards the centre. Proteins are also associated with this lipid matrix. The plasma membrane thus acts as a barrier to the passive movement of most molecules. Only fat-soluble molecules enter relatively freely, dissolving in the substance of the membrane and diffusing across. Since the interior of the membrane is largely lipid it contains little charge. Proteins are also associated with the membrane, intrinsic protein being incorporated into the bilipid layer and extrinsic protein being associated with the outer or inner face. Some proteins may span the membrane, providing water-filled pores. Small water-soluble molecules can enter the cell via these pores. Proteins may additionally provide more specialized mechanisms for the movement of certain substances, which form temporary complexes with them, so that these proteins have been given the name 'carrier' proteins. Proteins may also act as receptors for hormones and other chemical messengers, maintain the shape and structure of the membrane by, for example, combining with the underlying microfibrils and microfilaments and, in the form usually of glycoproteins, provide the determinants of the specificity of the cell. Membrane receptors may be associated with another group of proteins, namely enzymes responsible for energy-coupling reactions. TFTTF

The intracellular concentration of

1. Na^+ is the same as in the extracellular fluid.
2. K^+ is 5 mmol l^{-1}.
3. PO_4^{3-} is 95 mmol l^{-1}.
4. Cl^- is less than the extracellular fluid.
5. Na^+ is 12 mmol l^{-1}.

The intracellular fluid volume in man is approximately 30 l and the extracellular fluid volume is 15 l, comprising 12 l tissue or interstitial fluid and 3 l plasma. The electrolyte composition in the body compartments is kept constant. The main extracellular cation is sodium, present in concentrations of 145 mmol l^{-1}, plasma potassium concentration being only 5 mmol l^{-1}. The important anions quantitatively are chloride (110 mmol l^{-1}), bicarbonate (40 mmol l^{-1} in tissue fluid and 25 mmol l^{-1} in plasma) and, additionally in plasma, proteins (15 mmol l^{-1}), with low concentrations of sodium (12 mmol l^{-1}). The anions are mainly organic phosphate (90 mmol l^{-1}) and proteins (65 mmol l^{-1}). The observed distribution of cations results from an energy-requiring system in the cell membrane that pumps sodium out of the cell and potassium into it. It is this pump that maintains the osmotic equilibrium and hence the cell volume. If the pump fails, as in diseased cells, the high intracellular concentrations of proteins and other large molecules draw water in osmotically, causing the cell to swell. Chloride may or may not be passively distributed across the plasma membrane, depending on the cell type. Sodium chloride is generally ingested in excess of requirements, the excess being excreted in the kidney. Since sodium is an extracellular ion, increased plasma sodium (resulting from reduced excretion) leads to osmotic loss of water from the cells and increased blood volume, with resulting cardiovascular effects. If there is insufficient sodium taken in, for example to replace losses during excessive sweating, muscle cramps may result and possibly a drop in blood volume. FFTTT

MCQ 3

The sodium pump

1. pumps K^+ against the electrochemical gradient.
2. generally exchanges five Na^+ ions for one K^+ ion.
3. is inhibited by cyanide.
4. is inhibited by increasing intracellular Na^+.
5. uses much of the basal energy requirements.

The sodium–potassium pump moves sodium out of the cell and potassium into it, generally two potassium ions being transported inwards for three sodium ions outwards. The pump is envisaged as being a large phospholipid with ATPase activity and sodium and potassium binding sites. When the binding sites are occupied ATP is split and potassium is released intracellularly and sodium extracellularly. Potassium is therefore required outside to activate the pump, which is also stimulated by an increase in intracellular sodium. Cyanide blocks ATP synthesis and the pump cannot function without ATP. Much energy is required to maintain the pump; indeed, it has been estimated that up to 40–50 per cent of the basal oxygen consumption is required. The sodium pump is important in preventing cells from swelling. Living cells contain excess non-penetrating anions such as protein and phosphate. The greater osmotic pressure would mean that the cells would swell unless there were a mechanism to prevent it. The pump also contributes to the maintenance of normal intracellular ion concentration, necessary for many intracellular activities. Movement of sodium is also important for the transport of other substances which share a common carrier (e.g. glucose in the kidney and intestine). Finally, it contributes to the resting membrane potential and action potentials in nerve and muscle. The sodium pump works continuously as the membrane is permeable to sodium and potassium. It is some 50–100 times more permeable to potassium than to sodium. Because of this difference there is an unequal flow of positive ions into and out of the cell. The amount of potassium diffusing out is greater and because of the loss of positive ions a negative charge accumulates inside the cell, so that the inner face of the membrane is more negative than the outer. If the membrane were only permeable to potassium, then potassium ions would diffuse out until a sufficiently large potential difference built up for the rate of re-entry due to the electrical gradient to balance the loss down the concentration gradient. The potential difference across the cell would be the equilibrium potential for potassium (−90 mV). However, the diffusion of other ions contributes to the resting membrane potential (E) as given in the Goldman equation, which states that

$$E = \frac{RT}{F} \ln \frac{P_K[K_o]}{P_K[K_i]} + \frac{P_{Na}[Na_o]}{P_{Na}[Na_i]} + \frac{P_{Cl}[Cl_o]}{P_{Cl}[Cl_i]}$$

where P is the membrane permeability to a particular ion and the subscript 'o' refers to the outside concentration and 'i' to that inside. The resting membrane potential in most excitable cells is −70 to −80 mV and varies from tissue to tissue ranging from −9 to −100 mV. TFTFT

MCQ 4

Facilitated diffusion is characterized by

1. movement down an electrochemical gradient at a rate similar to simple diffusion.
2. a requirement for energy.
3. specificity.
4. saturation.
5. competition.

In facilitated diffusion a substance moves down its concentration gradient, as in simple diffusion, but at a greater rate. For example, glucose, which is important in metabolism, being a large polar molecule would not be expected to pass rapidly across the cell membrane. However, it moves in rapidly, far faster than hexahydric alcohol, which has a similar structure. Amino acids similarly enter cells at a rate greater than that produced by simple diffusion. It is postulated that this enhanced rate of transfer is made possible by combination of the substance with carrier molecules in the membrane, which were originally believed to shuttle to and fro across the membrane like ferries, but are now envisaged as being fixed—more like a toll bridge. The carrier hypothesis arose because of certain characteristics of facilitated diffusion. First, the process can be saturated—the rate of transfer increases linearly as the concentration increases and then at a certain point there is no further increase, as all the carrier sites are presumed to be occupied. Second, as with other processes involving biological binding of a molecule, there is specificity; carriers bind specifically with certain sugars and amino acids. The specificity is not absolute, so that certain related molecules can occupy the site, resulting in, thirdly, competition. Other molecules may bind irreversibly to the carrier causing, fourthly, inhibition of the transfer. Fifthly, counter-transport may be seen. For example, if glucose is occupying the carriers for transfer in one direction, a related sugar may move in the other direction. Finally, the process is very sensitive to changes in temperature, more so than would be expected for simple diffusion. A continuous supply of energy from metabolism is not required for facilitated diffusion. However, it is required for active transport, an example of which is the process whereby potassium is accumulated in the cell, while sodium concentrations are kept low. Characteristic of active transport is that it occurs against a concentration gradient. FFTTT

MCQ 5

Positive feedback

1. occurs in many biological systems.
2. prevents a rapid response.
3. occurs during the conduction of an impulse in nerve.
4. occurs during the menstrual cycle.
5. may be involved in the control of the anterior pituitary.

Central to physiology is the observation that the composition of fluid bathing the cells is maintained relatively constant, an observation made by Claude Bernard, who applied the term 'la fixité du milieu intérieur'. Canon extended this concept and called the integrated process 'homeostasis'. The respiratory system, circulatory system, digestive system and even the reproductive system contribute. Co-ordination of the internal organ systems and of the appropriate responses is effected via control systems. The essential feature of a control circuit is a closed-loop arrangement operating so that any disturbance of the controlled variable is automatically corrected. A simple control system comprises a sensor, which detects changes in the controlled variable and sends appropriate feedback signals to the controller. The controller compares this with the reference signal, and if the feedback signal differs from the reference signal the error causes the controller to initiate a compensatory response via the effector. Generally the function of the loop is to minimize deviation, the term negative feedback being used. More elaborate control systems exist, and also open-loop systems (i.e. with no feedback). Positive feedback systems occur only rarely in biological systems, as in such systems the greater the output the more the controlling system is activated. This, therefore, is an unstable system and can only exist if there is a natural end-point, as for example in the menstrual cycle, when maintenance of plasma oestrogen at a given concentration for a given length of time feeds back positively to produce the LH surge seen prior to ovulation. Positive feedback has an amplifying effect, generally producing a rapid response, as for example in the production of an action potential in a nerve. The stimulus causes a change in the membrane permeability, which results in an influx of sodium, partially depolarizing the membrane. This increases its permeability still further and permits more sodium to flow inward, until the peak of the impulse is reached.
FFTTT

The Fick principle

1. may be applied to the determination of blood flow to the brain using N_2O.
2. may be used to determine cardiac output.
3. applied to the determination of cardiac output requires determination of oxygen uptake.
4. applied to determination of cardiac output requires a sample of peripheral venous blood.
5. may be applied to the determination of cardiac output during hard exercise.

The Fick principle is in essence a reconfirmation of the law of conservation of mass; i.e., if a known amount of a substance is added to an unknown volume, then its concentration may be measured to estimate the volume. More strictly, the principle states that the amount of a substance taken up by an organ per unit time is equal to the concentration gradient of that substance across the organ times the blood flow. It can be used to determine the blood flow to an organ. The calculation can be performed using a substance which is either given off or taken up by the organ. The blood flow through the brain may be determined using N_2O. The subject breathes 15 per cent N_2O in an oxygen–nitrogen mixture for 10 min. As the partition coefficient of N_2O between brain and blood is unity, the amount of N_2O taken up by the brain can be calculated and hence the cerebral blood flow. The blood flow through the lung is essentially equal to that through the systemic circulation or the cardiac output. It may be determined from the oxygen uptake in the lungs and arteriovenous difference. If the oxygen uptake is 225 ml min^{-1}, the oxygen content of the arterial blood 190 ml l^{-1} and the venous content 140 ml, then

$$\text{Cardiac output} = \frac{225\ \text{ml min}^{-1}}{190\text{–}140\ \text{ml l}^{-1}} = \frac{225}{50}\ \text{l min}^{-1} = 4.5\ \text{l min}^{-1}$$

The quantity of oxygen absorbed by the lungs can easily be determined through the difference between inhaled and exhaled air. The arterial oxygen content can be determined from any systemic arterial blood sample because of the uniformity of systemic arterial blood. The main problem is the determination of the oxygen content of venous blood, since a mixed venous sample can only be obtained from the pulmonary artery. Catheterization can now be achieved using the Swan–Ganz flow-directed right heart catheter. This cardiac catheterization makes the technique too dangerous to use in exercise. Cardiac output may also be determined using an indicator dilution technique in which a given quantity of an indicator (dye, cold fluid, radioactive substance) is injected into an area of the cardiovascular system, where it is rapidly mixed with the blood flowing through and then sampled downstream. Cardiac output may also be determined by echocardiography, a technique based on reflection of an ultrasound wave, transthoracic impedance cardiography, in which flow in the aorta is determined by the degree to which the blood flow distorts an impedance wave from a high-frequency current, or by radionuclide imaging. TTTFF

MCQ 7

Plasma proteins

1. have a molecular weight of 30 000–85 000.
2. make a major contribution to the osmotic pressure of the plasma.
3. can contribute to nutrition.
4. are involved in the transport of calcium in the plasma.
5. contribute protection against loss of blood.

Plasma proteins form some 6.5–8 per cent by weight of plasma and have a molecular weight of 44 000–1 300 000. More than 85 per cent of plasma proteins (albumin, some globulins and fibrinogen) are synthesized in the liver, with the remaining 15 per cent being synthesized in the plasma cells. The relatively high viscosity of the plasma is due to the presence of plasma proteins, which are thus essential for the maintenance of blood pressure. Because of the high molecular weight of plasma proteins they make only a small contribution (25 mmHg, 3.33 kPa) to the total osmotic pressure, which is proportional to the number of osmotically active particles. However, the colloid osmotic pressure—that due to plasma colloids (oncotic pressure)—plays an important role in the exchange of fluids across the capillaries. Some of the plasma proteins are used in nutrition and in certain circumstances turnover of plasma protein can be so rapid that the dietary requirement of protein can be given intravenously in the form of plasma protein. There are a number of specific carrier proteins in plasma for both hormones, e.g. thyroxine-binding globulin and transcortin, and vitamins and also for trace elements, e.g. transferrin. There is also a less specific carrier function. Thus albumin can carry thyroxine, and all plasma proteins bind blood cations in a non-diffusible form. Because plasma proteins can also form salts by combining with acids and bases they contribute to the maintenance of a constant pH. Fibrinogen, an asymmetrical soluble protein, contributes to haemostasis, being converted to fibrin in the final stage of blood coagulation. The various globulins, referred to collectively as the immunoglobulins, form part of the defence system against bacterial antigens and foreign protein. FFTTT

MCQ 8

Lymph

1. contains no cells.
2. has an average protein content 10 per cent of that in plasma.
3. clots when left standing.
4. moves in the lymphatic system as a result, in part, of contraction of the smooth muscle in the vessel walls.
5. capillaries are found in the superficial layer of the skin and in the CNS.

The lymphatic system drains the excess tissue fluid, transports absorbed substances (particularly fats) away from the gastrointestinal tract and acts as a system for the circulation of competent lymphocytes for immunological processes. Although lymph is a filtrate, it has a surprisingly high protein content, reflecting the relatively high permeability of the capillaries to protein. The protein content of the lymph draining the legs is 10 per cent of that in plasma, while that for the liver is 80 per cent, giving an average protein content of 20 g l^{-1}. Lymph contains clotting factors and certain enzymes such as lipase and histaminase which enter the circulation via the lymphatics. The lymphatics arise as blind-ended capillaries found in all tissues except the superficial layers of the skin, the central nervous system (CNS) and the bone. The lymphatic capillaries join to form the lymphatic vessels, which empty into the large thoracic duct and the small right lymphatic. These then open into the great veins in the neck. Movement in these lymph vessels is brought about by the rhythmic contractions of the smooth muscle found in the vessel walls. As in the veins, flow is aided by the contraction of the surrounding skeletal muscle and there are also valves to prevent backflow. The normal lymph flow over 24 h is 2–4 l, but during muscular work the volume of lymph can be 10–15 times greater than at rest; hence the oedema which makes unused muscles stiff and painful after exercise. At intervals along the lymphatics are lymph nodes which act as filters so that minor infections can be localized. The so-called 'swollen glands' are a manifestation of this phenomenon. TFTTF

MCQ 9

Net movement of fluid from the capillaries into the interstitial space is

1. decreased by arteriolar constriction.
2. increased by venoconstriction.
3. decreased in dehydration.
4. decreased with a reduction in the circulating albumin concentration.
5. increased in hepatic failure.

Capillaries are exchange vessels taking oxygen and nutrients to the tissues and removing carbon dioxide and other end-products of metabolism from the tissues. Nutrient substances reach the cells through diffusion and bulk flow. Fluid leaves the capillaries via pores at the arterial end to bathe tissues and returns at the venous end. Starling first recognized the fact that whether fluid left the capillaries (filtration) or was reabsorbed depended on the balance of the two forces: the hydrostatic pressure tending to drive fluid out, and the colloid osmotic pressure tending to draw fluid into the capillary. If the hydrostatic pressure exceeds the osmotic pressure then net filtration will result, and vice versa. It is generally taken that the tissue pressure is positive, so that the net hydrostatic pressure at the arterial end of the capillaries is approximately 32 mmHg and at the venous end is 12 mmHg. The colloid pressure of the interstitial fluid is also small, so that the net colloid osmotic pressure is 25 mmHg. The formation of tissue fluid is thus influenced by factors altering the balance between these forces, enhanced tissue fluid formation resulting in oedema. Enhanced filtration will result from an increase in the hydrostatic pressure as seen in hypertension, vasodilatation as in exercise and hot weather, erect posture, increased blood volume produced by transfusion and increased pressure on the venous side as in cardiac failure. Hypotension as produced by haemorrhage and venous constriction will reduce loss of fluid from the capillaries. A decrease in the colloid osmotic pressure of plasma as in starvation and liver failure or accumulation of osmotically active substances in the interstitial fluid gives rise to enhanced filtration. Approximately 20 l of tissue fluid are formed per day and all but 10 per cent is reabsorbed. If the lymph vessels are blocked, as in inflammation, then oedema will result. TTFFT

Leucocytes

1. comprise granulocytes, lymphocytes, monocytes and thrombocytes.
2. are present in blood in concentrations of 4 000–10 000 × 10^6 l^{-1}.
3. are largely synthesized in lymphoid tissue.
4. are all capable of amoeboid movement.
5. have a circulation time of 90 days.

Leucocytes or white cells are nucleated colourless cells in the blood which circulate in concentrations of 4 000–10 000 × 10^6 l^{-1}. The three types of leucocytes are granulocytes (50–75 per cent), lymphocytes (20–40 per cent) and monocytes (2–8 per cent). Except for the lymphocytes, they are largely produced in the bone marrow and have a very short circulation time. Granulocytes, for example, spend a maximum of two days in the bloodstream. All white cells have the power of amoeboid movement and are attracted by many substances, including bacterial toxins and antibody–antigen complexes. They are able to engulf foreign material, a process known as phagocytosis. White cell counts may be performed microscopically, the cells first being diluted in a fluid containing acetic acid to haemolyse the cells and a stain for the nuclei. The nuclei of granulocytes are divided into lobes or segments (polymorphs), the lobulations being most pronounced in the neutrophils. These cells, which form the majority of the granulocytes, constitute one of the first lines of defence. They destroy bacterial cell debris and antibody-primed foreign matter. The second group of granulocytes comprises the eosinophils, the concentration of which increases in allergy. The numbers fluctuate through the day, apparently as a result of the changes in circulating glucocorticoids. The third group of granulocytes contains the basophils, which produce histamine and heparin. There are two types of lymphocytes: B-cells, which produce antibodies (after transformation into plasma cells), and T-cells, responsible for cell-mediated immunity. The monocytes may be transformed into tissue macrophages with high phagocytic activity. FTFTF

MCQ 11

Erythrocytes

1. have a life-span of 120 days.
2. contain a nucleus.
3. are present at a concentration of $5 \times 10^6\ l^{-1}$ blood.
4. are rigid biconcave discs 7.5 μm in diameter.
5. do not synthesize haemoglobin throughout their life-span.

The chief function of the red cells is to transport oxygen, a role accomplished by the haemoglobin they carry. Haemoglobin is not synthesized throughout the 120-day life-span of the cells as they lack the nuclei and metabolic machinery to synthesize the proteins. The cells also contain carbonic anhydrase, which aids in the transport of carbon dioxide. The cells are biconcave discs 7.5 μm in diameter and 2 μm thick. This shape allows the oxygen and carbon dioxide to diffuse rapidly into the cell interior. Red cells are capable of being easily and reversibly deformed to allow them to pass readily through the narrow curved capillaries. There are 5 million red cells per microlitre of blood, the counts being performed in the past by diluting the cells and counting them on a special slide or haemocytometer. In clinical practice red cells are now counted in an electronic counter. In adults the cells are produced in the red marrow of the flat bones and are destroyed by the macrophages found in the spleen, liver, bone marrow and lymph nodes, and the components recycled. When placed in a solution of reduced osmolality the red cells swell and finally burst. Using this property, fragility curves may be constructed. In a healthy individual 50 per cent of the cells haemolyse in a solution of 0.077M. Increased fragility is characteristic of acholuric jaundice. In a hypertonic medium red cells lose water and become crenated. In anticoagulated blood red cells settle out under the influence of gravity. In certain pathological conditions there is a tendency for rouleau formation so that this tendency for the cells to settle (erythrocyte sedimentation rate) is increased. TFFFT

Anaemia due to iron deficiency

1. is usually hypochromic and macrocytic.
2. may be a consequence of chronic haemorrhage.
3. is accompanied by a fall in the secretion of intrinsic factor.
4. is associated with the red cells circulating for about half the normal time.
5. gives rise to a megaloblastic bone marrow.

Of the 80 mmol of iron in the body 45 mmol is present as haemoglobin. About 0.45 mmol is released daily from the destruction of red blood cells and the breakdown of the haemoglobin they contain. There are losses of iron from the faeces, etc., so that to replace this amount some 20 μmol per day are required, which is provided by the daily intake of 90–180 μmol. Women require a greater intake to replace the menstrual losses (0.4 mmol) and an even larger supply during pregnancy. A deficiency of iron may thus arise from an increase in the needs of the body or from a lack of iron in the diet. It may also arise from poor absorption. Iron is generally absorbed in the ferrous form and is actively transported into the intestinal cells, where it becomes incorporated into ferritin, a protein–iron complex. Absorbed iron may then be released into the plasma, where it is bound to a specific protein transferrin. Deficiency of iron leads to a common deficiency disease: iron deficiency anaemia. This type of anaemia is called hypochromic, microcytic anaemia as the cells are pale through lack of haemoglobin and are small. This is shown by the low mean corpuscular haemoglobin concentration (haemoglobin in grams per 100 ml blood, divided by volume of red cells in millilitres per 100 ml blood). Anaemia may also result from acute or chronic blood loss, bone marrow aplasia, resulting for example from excessive exposure to X-rays, or from genetic causes as for example sickle-cell anaemia or thalassaemia. Pernicious anaemia is a condition which does not respond to iron replacement. It is a macrocytic anaemia in which the mean corpuscular volume (volume of red cells in millilitres per 100 ml blood, divided by number of red cells per 100 ml blood) is increased. The cells are irregular in shape and very fragile. The condition results from a deficiency of vitamin B_{12}, generally caused by failure of absorption which is dependent on intrinsic factor. Vitamin B_{12} is necessary for the maturation of erythrocytes, so the erythrocytes are immature. FTFFF

MCQ 13

Red cell production

1. occurs in the spleen of healthy adults.
2. requires vitamin D.
3. requires folic acid.
4. is affected by benzene derivatives.
5. is stimulated by erythropoietin.

Red cell formation or erythropoiesis occurs in the fetus in the spleen, but during the latter part of gestation and after birth it is limited to the bone marrow. By the age of 20 only the marrow of the flat bones produces red cells. They are formed from stem cells, which develop successively into proerythroblasts, erythroblasts, normoblasts and lastly into the reticulocytes, which have no nucleus but contain remnants of the cytoplasmic reticulum. Normally reticulocytes only account for 5–10 per cent of erythrocytes in healthy blood, but with accelerated erythropoiesis, as seen in haemolytic anaemia or at altitude, the percentage increases. Even under normal circumstances the rate of red cell production is high, with rapid mitosis, so that the process is susceptible to ionizing radiation and also to some drugs, especially the benzene derivatives. Red cell formation requires an adequate diet, providing sufficient iron and protein. Vitamin B_{12} (cyanocobalamin) is also necessary for red cell formation, as is another B vitamin, folic acid. There appear to be relatively large stores of vitamin B_{12}, but to obtain the vitamin from the diet intrinsic factor produced in the stomach is required for its absorption. While the supply of these factors influences red cell production, a glycoprotein hormone called erythropoietin exerts direct control. It is produced by the kidney and its release is stimulated by conditions which reduce the oxygen delivery to the tissues, as for example following haemorrhage or ascent to high altitude. FFTTT

MCQ 14

In the destruction of red cells

1. haem is split from globin.
2. biliverdin is reduced to bilirubin.
3. iron is lost in the urine.
4. bilirubin is conjugated to glucuronide.
5. the faeces are given their characteristic brown colour.

At the end of their life-span of 120 days it is not known if the cells haemolyse suddenly or gradually break down, but they are taken up into the reticuloendothelial system, where the phagocytes split down the haemoglobin molecule. This occurs in the liver, spleen and bone marrow, but the bruise marks caused by intracutaneous bleeding show that any tissue is capable of breaking down erythrocytes. Haemoglobin has a molecular weight of 68 000 and comprises four subunits, each subunit containing a haem moiety, which is an iron-containing porphyrin derivative which binds oxygen, and a polypeptide. The four polypeptides together form globin. There are two different types of polypeptide chain: α and β. The amino acid sequence of the polypeptides is genetically determined and variations in the sequence lead to production of different types of haemoglobin, some of them abnormal. The β chain is common to most haemoglobins. Adult haemoglobin has two β chains, but fetal haemoglobin has instead two γ chains, which differ from β chains by some 37 amino acids. Some of the alterations in polypeptide chains lead to severe disturbances of function, as for example in sickle-cell anaemia, in which haemoglobin molecules at low oxygen concentrations form fibre-like structures which cause the cells to assume sickle-like conformations and render them more likely to haemolyse. On breakdown of the red cells the globulin portion is split off and the components are re-used. The haem moiety is split open and the iron removed and transported in association with transferrin to the bone marrow to be reincorporated into the red cells. Any excess iron may be stored in the liver and other tissues. The haem portion is retrieved, the residual pigment being termed biliverdin. This is reduced to bilirubin, which being insoluble in water is bound to albumin during subsequent transport in blood. In the liver it is largely conjugated, mostly with glucuronic acid but a small amount with sulphate, and subsequently excreted in the bile. In the gastrointestinal tract there is further metabolism to stercobilinogen, which gives the faeces their characteristic brown colour. Of the bilirubin entering the duodenum each day about 10–20 per cent is not excreted but reabsorbed and recycled to the liver. Small amounts may be excreted via the kidney. If there is a blockage of the bile duct then the pigments are largely excreted via the kidney, so the faeces are pale and the urine very dark. TTFTT

MCQ 15

A woman of blood group O Rhesus-negative

1. has no antibodies to the ABO system in her blood.
2. has neither antigen A nor antigen B on the red cells.
3. could be a universal recipient.
4. has the most common blood type.
5. could not have a child that would be affected by Rhesus incompatibility.

If blood samples from two people are mixed, very often the erythrocytes clump, a process known as agglutination, as a result of an antibody–antigen reaction. The antigens (agglutinogens) are present on the red cell membrane as glycoproteins, while the antibodies (agglutinins) are present in the plasma. The different antigens give rise to the different blood groups of which the most important are A, B, AB and O. Groups A, B and AB have antigens A, B and AB respectively on the red cell surface, while group O has neither. During the first year of life antibodies are produced against antigens not present on the red cell. In giving a transfusion it is not significant if the donor plasma is not compatible, as it is rapidly diluted. However, if the antigens are not compatible then agglutination may occur, leading to haemolysis which can result in jaundice and possibly anaemia. Clumped red cells may also block the capillaries. Because no antigen is present on the cells of group O blood, such blood may be universally given (universal donor). Since there is no antibody in the plasma of type AB, then an AB patient can receive blood from any donor and hence is known as the universal recipient. With regard to inheritance alleles A and B are co-dominant. In European populations 46 per cent are group O, 42 per cent group A, about 9 per cent group B and the remainder AB. A further group of antigens, CDE, are found on the red cells, the most important being D. Blood containing D erythrocytes is called Rhesus-positive (85 per cent of the European population) and those without D are called Rhesus-negative. The antibodies of the Rhesus system do not occur unless the carrier has been exposed to Rhesus antigens. This is obviously significant on infusion, but more so in pregnancy, so if a Rhesus-negative mother has a Rhesus-positive fetus maternal antibody is induced. Usually the first child is not affected, but subsequent offspring will suffer with haemolytic disease. Current practice is to inject the mother with Rhesus-negative antibody within 36 h of childbirth to destroy any Rhesus-positive red cells which have invaded her circulation. FTFFF

Blood-clotting

1. requires calcium ions.
2. requires a sufficient dietary intake of vitamin E.
3. requires that prothrombin is precipitated.
4. cannot occur in haemophilia.
5. can be inhibited by heparin.

The initial description of blood-clotting was made early in the century by Morawitz and is still valid. Under the influence of thromboplastin, prothrombin is converted to thrombin, which in turn catalyses the conversion of fibrinogen to fibrin. Fibrinogen is a large rod-shaped protein of molecular weight 340 000 and during clotting it is split. The fibrin monomers thus formed become aligned and join together to form fibrin, a process requiring the presence of fibrinopeptide A and calcium. The polymer is gel-like in nature. Cross-links are formed and platelets are trapped within the network to give a strong fibrin clot. Platelets do not only contribute to clot formation, but also play a role immediately the blood vessel is damaged, allowing haemostasis within three minutes. On damage platelets stick to exposed layers of collagen. Adenosine diphosphate is released, which causes the platelets to become sticky so that they adhere to one another, forming a platelet plug. Vasoconstrictor substances such as 5-hydroxytryptamine and catecholamines are also released from platelets. Platelets in addition provide phospholipids, important in initiating the clotting process, being involved in the production of the active form of factor VII. This in turn catalyses the formation of factor X, which in turn is an important factor in the formation of thrombin. This pathway, named the extrinsic or extravascular system, can be activated within seconds. In the intrinsic or extravascular system the blood thrombokinase is activated within minutes. This system comprises a cascade of plasma proteins in which there is conversion of an inactive form to an active form by the preceding proteolytic enzyme in the sequence. Again the final agents responsible for thrombin formation are factors V and X and calcium. The process of clot formation is followed by clot retraction, which makes the clot more compact and stronger and aids in pulling the vessel walls closer together, thus helping to close the gap. A system also exists to dissolve the clot, a process known as fibrinolysis, which is catalysed by fibrinolysin (plasmin). This protease splits soluble peptides from fibrin. It is formed from an active precursor in a manner analogous to the formation of thrombin. Another factor important in blood-clotting is vitamin K. Bile salts are required for the absorption of this vitamin from the gastrointestinal tract, so that liver disease or deficiency of fat absorption may lead to bleeding problems. Blood does not normally clot in blood vessels as an anti-clotting system exists, but will do so *in vitro*. This can be prevented by chelating agents such as EDTA, which remove calcium from the plasma and heparin, which prevents activation of factor IX. TFFFT

MCQ 17

An action potential

1. is initiated by membrane depolarization.
2. has a magnitude depending on the stimulus strength.
3. is accompanied by an increase in the electrical resistance of the membrane.
4. can travel in either direction.
5. has a peak which depends on the external sodium concentration.

The passage of an action potential along a nerve or muscle is the means by which a signal is conducted in these excitable tissues. The action potential is initiated by membrane depolarization. Once the membrane is depolarized to the threshold potential—a change of 20 mV relative to the resting potential (which is close to the equilibrium potential of potassium)—then there is an increase in the membrane conductance to sodium ions, which flow inwards. If the potential changes are measured using an intracellular electrode, then this is recorded as the upstroke or rising phase of the action potential. This is followed by the overshoot, which is a short, positive phase with a peak of approximately 30 mV, at which point the potential becomes much closer to that of the sodium equilibrium potential. The rapid rising phase of the action potential results from the positive feedback relationship between membrane depolarization and sodium permeability: the greater the influx of sodium the greater the depolarization and the greater the increase in sodium conductance. This explains the all-or-nothing nature of the action potential. The amplitude of the action potential is independent of the initiating event and the response cannot be summed. The next phase is repolarization. As the membrane potential becomes more positive, the sodium conductance falls to resting values. The repolarization is speeded up as the potassium gates in the membrane open later than those for sodium and stay open longer. Outward flux of sodium slows the rise of the potential, produces a

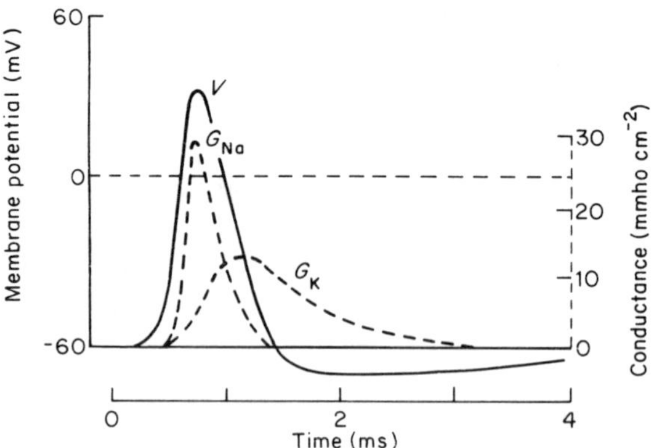

Figure The time-course of the computed conductances during an action potential in a squid giant axon. V is the voltage across the membrane and G_{Na} and G_K are the conductances of sodium and potassium, respectively. The equilibrium potentials for sodium and potassium are +70 mV and −90 mV, respectively (after Hodgkin and Huxley).

fall in the potential and even briefly reverses it. This is called a hyperpolarizing after-potential. In some types of action potential, as in the muscle, the rapid repolarization is followed by a slowly changing potential, the depolarizing after-potential. The action potential lasts about 1 ms in mammalian nerve and 10 ms in skeletal muscle. These membrane changes mean that for a brief period after the action potential (about 1 ms in the nerve) the fibre is unresponsive to a depolarizing current (see MCQ 20). The direction of conduction of an action potential depends on the anatomical arrangement of the system: theoretically an action potential could travel in either direction. TFFTT

MCQ 18

The velocity of conduction of an action potential in nerves

1. is decreased with increasing fibre diameter.
2. is slower in myelinated nerves.
3. is increased by decreasing temperature.
4. is 20–100 m s^{-1} in somatic motor nerve fibres.
5. can be reduced by decreasing the external sodium concentration.

The depolarization produced on stimulating a nerve causes a flow of current (local current) between the resting portion of the fibre and the depolarized region. Current flows out of the membrane in the resting region, through the surrounding medium to the area of depolarization and then back through the membrane, the circuit being completed by conduction through the core of the nerve fibre. This flow of current depolarizes the adjacent area of the membrane, bringing it to the threshold so that an action potential is generated in a new region. As the region excited becomes refractory, the propagation of the potential is usually unidirectional. Since the spread of current depends on the flow of sodium, anything which reduces it will result in a reduction of the action potential amplitude and hence of the longitudinal potential. Therefore decreasing the external sodium concentration will reduce conduction velocity. The conduction velocity is also determined by the resistance of the axoplasm. Bigger axons, which have a larger cross-sectional area, have small internal resistances. Thus the conduction velocity increases as the square root of the fibre diameter. Another factor which influences conduction velocity is myelination. Some nerve fibres are surrounded by a sheath of myelin, interrupted at intervals by nodes of Ranvier. The sheath lends high electrical resistance to the membrane so that an action potential at one node spreads to the adjacent nodes electro-ionically. This jumping from one node to another occurs at the speed of light and is called saltatory conduction. Myelination also reduces the membrane capacity so that less current is needed to charge the membrane between nodes. The speed of conduction in myelinated fibres is approximately 10 m s^{-1} in B fibres (average diameter 3 μm) to about 100 m s^{-1} in A fibres (15 μm average diameter). In invertebrates, where only non-myelinated fibres are found, high conduction velocities are only achieved by increasing the fibre diameter to 1 mm. FFFTT

MCQ 19

In stimulating and/or recording from a nerve trunk

1. extracellular electrodes may be used to record the resting potential.
2. extracellular electrodes always give a monophasic action potential.
3. the period between the stimulus and the response allows estimation of the conduction velocity.
4. recording of the stimulus artefact is prevented by crushing the nerve between the recording electrodes.
5. the shape of the compound action potential is partly dependent on the stimulus strength.

Action potentials may be recorded using a pair of electrodes. If one electrode is inside the cell and the other is outside, the potential across the membrane is measured. The amplified potential changes can be displayed on a cathode-ray oscilloscope as a monophasic action potential. Usually it is more convenient to record with both electrodes outside the nerve—an extracellular recording. If both electrodes are over an intact part of the nerve and separated by sufficient distance, then a biphasic action potential is recorded. With surface electrodes a comparison is obtained of the potential between two regions of the cell membrane surface. The potential difference is recorded at one electrode and then a mirror image is seen as the propagated action potential reaches the second electrode. If the nerve under the second electrode is cut or crushed the impulse never reaches this point and a monophasic recording is obtained. If a frog's

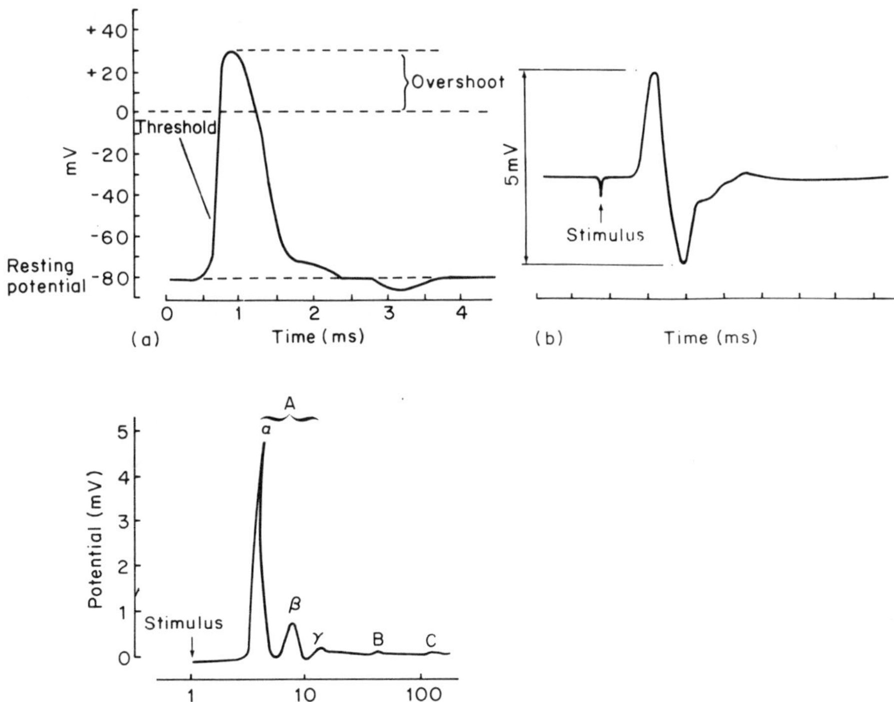

Figure Action potential with (a) recording electrode inside the cell, (b) electrodes on the surface of the cell and showing (c) a compound action potential recorded from the sciatic nerve of a frog. The fibre types are indicated.

sciatic nerve is strongly stimulated and a monophasic recording of the potential is made some distance from the stimulating electrode, three groups of spikes can be distinguished—A, B and C—the A wave being further subdivided into α and β. Each part of this complex represents different groups of nerve fibres, each with different rates of conduction, A being the fastest (conduction velocity 70–120 m s^{-1}) and C the slowest (0.5–2.0 m s^{-1}). The equivalent classification to A, B and C for motor nerves is I, II and III for sensory nerves. The stimulus strength does not influence the rate of conduction, the small fibres always conducting more slowly, but it does affect the initiation of an action potential. Smaller fibres require a greater stimulus strength and duration. FFTFT

MCQ 20

Membrane excitability is increased by

1. hypocalcaemia.
2. mild hypoxia.
3. long-term exposure to ouabain (a sodium pump inhibitor).
4. local anaesthetics.
5. small increases in extracellular K^+.

The excitability of a membrane is a measure of the ease with which an action potential can be generated. During the period of increased sodium conductance associated with the action potential the membrane cannot be stimulated again, which is the reason why action potentials cannot summate. The unresponsive period is the absolute refractory period. As the sodium conductance returns to normal the membrane excitability returns, but for a short period—namely the relative refractory period—a larger stimulus is required to produce what is in fact a smaller action potential. After the relative refractory period the membrane is partially depolarized and the potential is closer to threshold, so that there is increased excitability. This is the supernormal phase, which is followed by the subnormal period when the membrane is hyperpolarized away from threshold before the resting potential is restored. Calcium acts as a membrane stabilizer, so that if there is a reduction in the calcium concentration there is an increase in the magnitude of the surface potential due to the negative protein charges in the membrane. This would affect the sodium conductance and hence the inward current. During hypoxia the ionic concentration gradients cannot be maintained and the resting potential decreases, which may be associated with spontaneous firing. Ouabain produces blockade of the electrogenic component of the sodium–potassium pump, so that relatively prolonged exposure to this drug would allow the gradients for sodium and potassium to run down and the resting potential and overshoot to approach zero. As a result of the effect of depolarization on the sodium channels the sodium system would inactivate and the excitability would disappear. Local anaesthetics would block sodium channels and hence sodium entry during the action potential, reducing excitability. Increasing external potassium ion concentrations would depolarize the resting potential and hence increase excitability. TTFFT

MCQ 21

In the synaptic release of a neurotransmitter

1. there is a decrease in the conductance of the presynaptic membrane.
2. the quantity released depends on the size and duration of the action potential.
3. there is depolarization of the presynaptic membrane.
4. magnesium causes quantal release.
5. calcium enters the presynaptic terminal.

A synapse is a junction between neurones, with information being passed from the presynaptic to the postsynaptic neurone either by release of a transmitter (chemical synapse) or by flow of current (electrical synapse). Neurotransmitters are thought to include acetylcholine, noradrenaline, dopamine, γ-aminobutyric acid, glutamic acid, 5-hydroxytryptamine and histamine. Although terminals can synthesize some substances such as acetylcholine they are incapable of protein synthesis and so require material to be transported from the cell body. Transmitters are contained mainly in vesicles and are released from the presynaptic terminal in quanta. The release of quanta is a matter of probability, so that at any time quanta may be released. The probability is increased a thousandfold with an action potential which represents a transient depolarization of the membrane. The mechanism of production of an action potential is the same as in the nerve fibre. The quantity of transmitter released depends on the amplitude and duration of the action potential. Release of quanta can also depend on the membrane potential of the presynaptic ending, increasing as the membrane potential becomes more positive and decreasing as it becomes more negative. Depolarization of the membrane (about 30 mV) opens the calcium channels so that calcium flows into the terminal down the concentration gradient, allowing exocytosis of the vesicle to occur. The presence of calcium is a prerequisite of quantal release in response to a presynaptic action potential. It has been calculated that four calcium ions are needed to burst one acetylcholine vesicle. Addition of magnesium ions has a similar effect to the removal of calcium ions, so it appears that magnesium ions compete with calcium ions for the channels but cannot substitute for calcium in transmitter release. FTTFT

MCQ 22

Termination of transmitter action may involve

1. diffusion away from the synaptic cleft.
2. enzymic breakdown at the muscle endplate.
3. removal by the postsynaptic membrane.
4. transport away from the synaptic cleft by calcium.
5. re-uptake into the presynaptic terminal.

Important for the functioning of the nervous system is the fact that the transmitter acts at the postsynaptic membrane or endplate for a very limited period of time (about 1.2 ms). Very efficient methods of inactivating transmitters must therefore exist. Acetylcholine is broken down into choline and acetic acid by the enzyme acetylcholinesterase, which is found in large amounts at the motor endplate. The components of acetylcholine are then taken back into the presynaptic terminal and the transmitter resynthesized. Acetylcholine also diffuses away from the synaptic cleft and may be broken down by the so-called non-specific or pseudocholinesterases in the plasma. There are also enzymes which destroy the catecholamines, namely monoamine oxidase (MAO) and catechol-*O*-methyltransferases. However, they seem to play little role in the termination of transmitter activity, as neither has a marked effect on the response to the administration of noradrenaline. The chief mechanism in terminating the action of noradrenaline is re-uptake of the transmitter into the presynaptic terminals. This mechanism is dependent on an intact supply of energy via aerobic or anaerobic pathways and on the surrounding sodium ion concentration and hence is active. It is assumed that other catecholamines are similarly handled. This mechanism has the additional advantage that stores of catecholamines are not depleted. Glutamate is also actively taken up. TTFFT

MCQ 23

An excitatory postsynaptic potential

1. consists of hypopolarization of the postsynaptic membrane.
2. is produced by a simultaneous increase in membrane permeability to sodium and potassium ions.
3. is followed by a refractory period.
4. may summate with other EPSPs.
5. contributes to the activity of motor neurones in the stretch reflex.

After release from the synaptic terminal the transmitter diffuses across the synaptic cleft to the postsynaptic or subsynaptic membrane. Here it binds to receptors, the charge produced depending on the nature of the synapse and the structure of the postsynaptic membrane. At an excitatory synapse there is a simultaneous opening of the pores for sodium and potassium ions, so that there is hypopolarization of the postsynaptic membrane, and the potential is brought closer to the threshold—an excitatory postsynaptic potential (EPSP). At an inhibitory synapse the membrane potential is moved further away from the threshold value (IPSP). When recorded at an excitatory synapse, the depolarization curve rises rapidly over 2 ms and decays slowly over 10–15 ms. There is no refractory period, which is important as generally a single EPSP does not produce a sufficient change in potential for an action potential. EPSPs are required to summate to evoke an action potential. Since the transmitter is released in quanta, the EPSP produced is graded in size depending on the number of quanta reaching the postsynaptic receptors. The additive effect of varying inputs can be achieved by temporal or spatial summation. Volleys of impulses do not arrive synchronously, but if the second EPSP arrives before the first has died away they can summate—temporal summation. When activity arrives via several individual presynaptic elements, and the transmitter is released almost at the same time, the individual synaptic potentials summate to make a large potential change—spatial summation. The effect of the EPSP can be reduced by IPSPs, which can also summate. Axosomatic and axodendritic synapses occur also. The axon hillock has a lower threshold for triggering an action potential than the dendrites, so is important for initiating an impulse. In addition it has been observed that the excitability of the motor neurone is greater the smaller the neurone. TTFTF

MCQ 24

Inhibition at synapses can result from

1. hyperpolarization of the postsynaptic membrane.
2. increased chloride conductance of postsynaptic membranes.
3. release of γ-aminobutyric acid from the presynaptic terminal.
4. activation of a Renshaw cell.
5. presynaptic inhibition with reduced release of transmitter from the presynaptic terminal.

Not all synapses within the central nervous system are excitatory: some when activated inhibit the postsynaptic neurone. At an inhibitory synapse hyperpolarization of the postsynaptic membrane or an inhibitory postsynaptic potential (IPSP) occurs. Inhibition can also be produced by increased chloride conductance with no hypopolarization. This shunts the positive currents, so that no change in membrane potential occurs at the soma. Established transmitters include acetylcholine, noradrenaline, γ-aminobutyric acid and glutamate. While acetylcholine is a transmitter at excitatory synapses, the inhibitory transmitter at motor neurones is glycine. The amino acid derivative γ-aminobutyric acid is also an inhibitory neurotransmitter. Inhibition may also occur presynaptically, a phenomenon which protects the system against the barrage of sensory inputs received. Presynaptic inhibition is particularly important at lower levels of the brain and spinal cord, where it produces a widespread suppressive effect. Presynaptic inhibition entails the reduced release of neurotransmitter from the presynaptic terminal, resulting from a reduction in the size of the invading action potential. This is achieved by a reduction in the number of voltage-dependent calcium channels that are opened. Inhibition may also be produced via inhibitory interneurones; those acting on motor neurones are called Renshaw cells. These cells have to be excited by a large number of motor neurones before they fire, acetylcholine being the excitatory neurotransmitter. The Renshaw cell then has an inhibitory effect on the motor neurones, producing an IPSP through the release of glycine. This is an example of feedback inhibition, also called recurrent inhibition. TTTTT

MCQ 25

In junctional transmission

1. acetylcholine is involved in the excitation of muscle at the anterior horn.
2. strychnine blocks the effects of inhibitory synaptic potentials.
3. curare blocks the liberation of acetylcholine.
4. the bacterium *Clostridium botulinum* releases a toxin that blocks transmission by binding at the acetylcholine receptor site.
5. atropine enhances the action of acetylcholinesterase.

Acetylcholine is a major transmitter within the nervous system and at the neuromuscular junction. The receptors for acetylcholine are specific proteins found in the postsynaptic membrane in great concentration. The conformation alters on binding of acetylcholine, which results in changes in the ionic permeability of the membrane. There are two types of acetylcholine receptors, originally characterized by Dale on the basis of the response to nicotine and muscarine. Both these alkaloids can mimic the action of acetylcholine, but at different sites. Nicotinic receptors (i.e. responsive to nicotine) are found at the neuromuscular junction and in sympathetic and parasympathetic ganglia. Muscarinic receptors are found in the tissues responding to parasympathetic stimulation, for example the heart and smooth muscle of the gastrointestinal tract. Acetylcholine is not involved in transmission at the anterior horn. Synaptic transmission may be blocked or modified by neuropharmacological agents acting at any one of the various steps involved. The nerve impulse may be blocked by local anaesthetics and tetrodotoxin and the subsequent calcium entry into the presynaptic terminal by a number of agents, including heavy metals and propranolol. Release of acetylcholine may be blocked by substances such as tetanus toxin and botulinis toxin produced by the bacterium *Clostridium botulinum*. The latter is responsible for a type of food poisoning and is extremely potent, being lethal in a dose of 0.1 μg. Neuromuscular transmission may also be disturbed by inhibition of acetylcholinesterase by neostigmine, physostigmine and some organophosphates which are the chief ingredients of some pesticides and of nerve gases. Choline uptake into the nerve ending is inhibited by hemicholinium. If acetylcholine is not destroyed its activity is prolonged and depolarization is maintained. Finally there are drugs which bind to the acetylcholine receptor. These include the deadly poison curare and a number of snake venoms. Atropine binds to muscarinic receptors. FTTFF

At the neuromuscular junction

1. there are marked postsynaptic folds.
2. there is a synaptic cleft of 10–50 nm.
3. an action potential in the nerve is unable to generate an action potential in a muscle unless it summates the arrival of other action potentials.
4. succinylcholine causes depolarization of the postsynaptic membrane.
5. muscle is insensitive to acetylcholine after section of its nerve.

Axons of the motor neurones leaving via the central horns synapse with the skeletal muscle fibres. These junctions are the neuromuscular endplates. The motor nerve terminals lose their myelin sheath and divide into a spray of non-myelinated branches about 1 μm in diameter which lie on indentations in the surface of the muscle cell. The axon terminal is separated from the muscle fibre membrane or sarcolemma by a gap of 10–50 nm in width. The postsynaptic membrane has marked folds, increasing the surface area of contact. In the presynaptic nerve terminal is found an orderly arrangement of synaptic vesicles 40–200 nm in diameter which contain acetylcholine. In the postsynaptic membrane are high concentrations of particles in the edges of the synaptic folds and these are believed to be the receptors. There is a delay (synaptic delay) between the nerve impulse reaching the neuromuscular junction and the first indication of postsynaptic depolarization. In the frog neuromuscular junction at room temperature this can be 0.5 ms. This is largely due to the time for Ca^{2+} to enter the presynaptic terminal and effect fusion of the synaptic vesicles with the axolemma. After section of the nerve, the muscle becomes hypersensitive to acetylcholine, responding to acetylcholine over its whole surface. The hypersensitivity results partly from the lack of neuromuscular activity and from the removal of neuroendocrine synaptic factors that affect metabolism of the muscle fibre. Succinylcholine mimics the action of acetylcholine, but has a prolonged effect. Stimulation of muscle contraction is thus blocked by the resulting depolarization of the motor endplate. It is thus called a depolarizing muscle relaxant. The neuromuscular junction represents an excitatory synapse, but no integration occurs here; instead each nerve impulse arriving generates an action potential in the muscle fibres it supplies. TTFTF

MCQ 27

Common to skeletal, cardiac and smooth muscle are

1. a stable resting membrane potential.
2. an action potential of duration 200 ms.
3. electrical connections between cells through gap junctions.
4. a requirement for calcium for contraction.
5. an involvement of actin and myosin in contraction.

Skeletal muscle has a stable membrane potential, as do atrial and ventricular muscle cells, with a resting potential of approximately −80 mV. In contrast, the cells of the sinoatrial node, the atrioventricular node and the Purkinje fibres do not maintain a steady membrane potential; instead the cells slowly depolarize, giving a pacemaker potential. The cells depolarize until a threshold is reached and an action potential generated. These cells therefore contract in a spontaneous rhythmic manner, the rate depending on the rate at which the membrane potential approaches threshold. Important in these changes appears to be a slowly decaying potassium current. Smooth muscle, too, has no true resting potential. The potential recorded is high when the the tissue is active and lower when it is inactive. Spontaneous electrical activity is seen, some being of the pacemaker type. For example, cells with a spontaneous frequency are located in the stomach, waves of activity moving down the body of the stomach along the great curvature. If smooth muscle is stretched there is an increase in the frequency of the spikes of electrical activity. The plasma membrane of skeletal muscle of the rapid twitch type has functional properties similar to those of an unmyelinated nerve. The resting membrane potential is about −90 mV and the action potential lasts about 2 ms, which is extremely brief compared to the duration of the ensuing contraction. In contrast, the action potential in cardiac muscle is complex and lasts over 200 ms, much closer in duration to that of the contraction, so that the heart muscle cannot be tetanized. The cells of skeletal muscle are electrically independent, co-ordinated contraction resulting from the fact that many muscle fibres are supplied by the same motor neurone. There are between the cells of smooth and cardiac muscle what are known as gap junctions where adjacent cells join, producing a point of electrical contact so that impulses are propagated from one cell to another. If a preparation of ventricular muscle is stimulated the whole preparation responds before repolarization begins. The mechanisms of contraction are similar in different types of muscle. Intracellular calcium is important in contraction, but in smooth muscle is supplied from external sites and is pumped into the cell. Actin and myosin form the basis of the contractile response but in smooth muscle there is evidence that calcium may act directly with the myosin cross-linkages. FFFTT

Smooth muscle

1. when stretched may depolarize and contract.
2. always possesses inherent activity, even when the nerve supply is cut.
3. may have pacemaker cells.
4. is always depolarized on application of noradrenaline.
5. occurs in the cardiovascular system, the gastrointestinal tract and the reproductive system.

Smooth muscle is found in many organs and tissues throughout the body and its organization varies to suit the differing physiological requirements. It usually forms part of the wall of tubular or sac-like organs. The cells themselves are 5–10 μm in diameter and 200–500 μm long and are grouped in a parallel array. In organs such as the gastrointestinal tract and the uterus there are several layers of smooth muscle. In contrast it may be only one cell thick in the very small blood vessels. Smooth muscle may also be modified to form sphincters in the gastrointestinal tract, blood vessels and urinary bladder. It is found in many other sites, including the respiratory passages and the skin. Smooth muscle is the effector organ of the autonomic nervous system. Multi-unit smooth muscle has innervation similar to that of skeletal muscle, except that there may be overlapping innervation. Little or no spontaneous or rhythmic activity is present, the nerve supply being excitatory. Multiple action potentials in the nerve are required to depolarize the muscle cell sufficiently to reach the threshold for an action potential. Circulating hormones may also initiate contraction of the muscle. Examples of multi-unit smooth muscle are the iris and the piloerector muscle. Single-unit muscle, on the other hand, contracts spontaneously and rhythmically in response to changes in the local chemical environment, or to stretch. This activity is modified by impulses from the autonomic nervous system, which either inhibits or excites further activity. In general, activity in the sympathetic nervous system prepares the body for an active response to an adverse environment, while the parasympathetic fibres control internal influences under quiet conditions. Noradrenaline does not always have the same effect on smooth muscle. While it produces hyperpolarization and hence relaxation in smooth muscle of the intestine, it produces depolarization and contraction in the walls of blood vessels. TFTFF

MCQ 29

In skeletal muscle

1. the thick filaments comprise myosin.
2. actin and tropomyosin are found in the thin filaments.
3. the size of the H band decreases during contraction.
4. the A band contains only thin filaments.
5. the A bands shorten, but not the I bands.

When viewed under a light microscope, the most striking and characteristic feature of skeletal muscle is the presence of light and dark bands, which give rise to the term striated muscle. This banding is confined to the myofibrils, of which there are several hundred or in some cases several thousand per fibre. Each fibre represents a multinucleate cell, which can be 1–50 mm long and 10–100 μm in diameter. The cell membrane is called the sarcolemma and the cytoplasm the sarcoplasm. One unit of a repeating pattern of striations on a myofibril is called a sarcomere and is composed of thick myosin filaments and thin actin filaments, which overlap to produce an A band. In the centre of the A band is a lighter H zone, which corresponds to the space between the ends of the thin filaments and in which there is no overlap between the thick and thin filaments. The region in which the thin filaments alone are found is called the I band, across which runs the Z line, formed by strands into which the thin filaments insert. The Z lines repeat at intervals of 1.5–2.5 μm, depending on the degree of shortening or stretching of the myofibril. According to the sliding filament theory of Huxley the muscle shortens as a result of the relative movement of the thick and thin filaments. The sliding movement is produced by the oscillating movements of cross-links formed when bridges on the myosin filaments briefly join onto the actin filaments. On contraction the Z lines come closer together, reducing the length of the I bands and also the H bands which are finally obliterated. Tropomyosin found in association with actin contributes to the contraction mechanism. Other structures important in contraction are the components of the dual membrane system. The first system is open-ended and formed by the sarcolemma, which penetrates the muscle by invaginations forming transverse tubules. The other system, the sarcoplasmic reticulum, is a longitudinal system, which makes membranous connections with the transverse tubular system

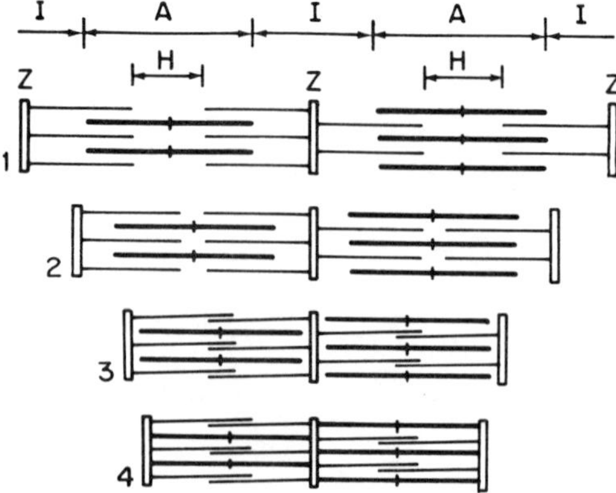

Figure Diagram illustrating the filaments sliding past each other as the muscle gradually contracts from 1 to 4 (after Huxley and Hanson).

to form the triads. These triads comprise two lateral sacs of the sarcoplasmic reticulum and one transverse tubule. Action potentials are rapidly conducted along the membrane system to the lateral sacs, where the stored calcium is released. TTTFF

MCQ 30

In skeletal muscle ATP

1. can be formed from oxidative phosphorylation in the mitochondria.
2. can be formed by substrate phosphorylation during glycolysis.
3. is slowly formed from creatine phosphate.
4. provides energy for cross-bridge movement.
5. splitting breaks the link between actin and myosin.

Hydrolysis of ATP provides the energy for skeletal muscle contraction. Muscle cells contain similar quantities of ATP to other cells, but this is not enough to provide the energy required for contraction. However, ATP can rapidly be replenished by the transfer of phosphate from creatine phosphate to ADP. The process is so efficient that the concentration of ATP changes little until the creatine phosphate stores are almost depleted. While creatine phosphate provides a rapid means of forming ATP, there are other ways of maintaining ATP concentrations if exercise continues. In moderate exercise ATP is produced aerobically by oxidative phosphorylation, fat being the preferred fuel, although glucose and glycogen also contribute. As the intensity of the exercise increases, the supply of oxygen and nutrients becomes inadequate so the supply of ATP is maintained by glycolysis. Not only does this process proceed in the absence of oxygen, but it also produces ATP rapidly. However, in contrast to oxidative phosphorylation, in which 36 molecules of ATP are produced for each molecule of glucose, only two molecules of ATP per glucose are produced. There are several ways in which ATP can contribute to muscle contraction. First, it provides energy for the movement of cross-bridges between actin and myosin, which produces the sliding of filaments underlying muscle contraction. ATP is hydrolysed, producing a high-energy form of myosin, M.ADP.P, which forms the AM.ADP.PI complex. A change in the conformation of this complex is believed to generate active tension. Second, the binding of another ATP molecule to myosin weakens the bond between actin and myosin and allows dissociation of actin and myosin, so that the cyclical formation of bridges is possible. A third role is in relaxation. Contraction is terminated when the calcium is removed from troponin, thereby blocking the myosin binding site on the actin. The energy to pump calcium back into the reticulum comes from ATP. TTFTF

MCQ 31

During muscle contraction calcium

1. is released from the plasma membrane directly, producing contractile activity.
2. triggers contraction by binding to tropomyosin.
3. uptake by the sarcoplasmic reticulum is associated with ATP hydrolysis.
4. uptake by the sarcoplasmic reticulum allows relaxation.
5. concentration in the cytosol surrounding the filaments is lower than in the resting muscle.

Calcium plays an important role in muscle contraction, which is induced by an action potential propagated over the surface of the muscle membrane and then into the muscle fibre along the transverse tubule. The spread of excitation in the T tubule leads to release of calcium from the sarcoplasmic reticulum. The sarcoplasmic reticulum is composed of an elaborate internal membrane system which surrounds the myofibrils and runs close to the system of transverse tubules. Release of calcium results in an elevation in concentrations, which under resting conditions are below 10^{-7} M, to about 10^{-5} M. Under conditions of low calcium concentration the formation of cross-linkages between actin and myosin which forms the basis of muscle contraction is inhibited. On release of calcium from the terminal cisternae, some binds to troponin located on the thin actin filaments. This causes the tropomyosin to move to one side, exposing the binding sites and allowing the cross-bridges to bind to the thin filaments. The calcium is rapidly taken up by the sarcoplasmic reticulum so that the concentration falls again, the blocking action of tropomyosin is restored and relaxation ensues. Calcium uptake involves active transport, the energy for which is supplied by the hydrolysis of ATP. Calcium is taken up through the longitudinal sarcoplasmic reticulum to be stored in the terminal cisternae. Calcium is stored in the internal membrane system, where it is bound to a number of different calcium-binding proteins. FFTTF

MCQ 32

In muscle fatigue there is

1. failure of conduction in the motor nerve.
2. reduction in neurotransmitter release.
3. an abnormal spread of action potential across the muscle membrane.
4. a difference between the effect produced by tetanic stimulation of the isolated muscle and the effect of the natural response.
5. depletion of ATP in the muscle fibres.

With repeated stimulation of the isolated muscle preparation fatigue is seen. The latent period is increased, the contraction becomes smaller and more prolonged, with slow relaxation, and the muscle may not return to its original length. Finally the muscle can no longer contract, a state similar to rigor mortis, in which all the cross-bridges are irreversibly locked. It may also be seen *in vivo*, as for example in the contraction of the adductor pollicis in response to repetitive maximum shocks to the ulnar nerve. In these last studies it was shown that the action potential in the muscle fibres was of normal amplitude and frequency, indicating that the contractile mechanism was involved and that the neuromuscular transmission was unimpaired. It has been generally assumed that failure of contraction is due to depletion of energy stores, in particular ATP and creatine phosphate, but recently evidence has been provided that impairment of excitation–contraction coupling may be important. Susceptibility to fatigue depends on the type of muscle fibre. Three types of muscle fibre can be distinguished on the basis of their ability to generate ATP and the speed of contraction. These are slow-twitch oxidative, fast-twitch oxidative and fast-twitch glycolytic. The latter have a high myosin ATPase activity, a low oxidative capacity and high glycolytic components. The speed of contraction is rapid and fatigue also has a rapid onset. The problems associated with fatigue in the intact animal during exercise are more complex. Muscle is affected by hormones influencing metabolism and by circulating and many other factors. There is also central fatigue, in which the central nervous system cannot supply the required train of impulses to maintain the exertion at a peak. There is also transmission fatigue, in which the motor endings become depleted of acetylcholine. Delay in transmission at synapses may also contribute to fatigue associated with boredom or staleness. This type of fatigue has a psychological basis. FTFTT

MCQ 33

The tension in skeletal muscle

1. is called passive if generated by stretching the elastic tissues.
2. at rest involves ATP hydrolysis.
3. always increases as muscle length increases.
4. during active contraction is maximal at the normal muscle length.
5. is greater during tetanus than a single twitch.

In parallel with the contractile apparatus of the muscle is the so-called parallel elastic component provided by connective tissue. If a muscle is allowed to contract with both ends fixed, so that there is little or no change in the muscle length, the response is termed isometric. The tension increases as the contractile elements shorten, pulling and elongating the elastic elements. In an isotonic contraction the tension is held, the series elastic elements are passively stretched by the weight lifted and an equilibrium is established between the tension and the load at any one point, so that the shortening of the contractile elements raises the load. Voluntary movements are a mixture of isotonic and isometric contractions. The characteristics of these types of muscle contraction may conveniently be studied in an isotonic nerve – muscle preparation, which in its simplest form consists of a muscle such as the frog sciatic with its motor nerve still attached. The nerve is stimulated electrically and the contraction recorded using an electronic transducer, which converts mechanical energy to an electrical signal which can produce a record on an electronic recorder or an oscilloscope. If the contraction is recorded under isotonic conditions there is a short latent period on stimulation, followed by a brief contraction phase and then relaxation. If a second stimulus is applied before the end of relaxation then the second response is greater than the first. If a series of stimuli are given then the tension increases to a maximum, known as tetanus. In tetanus there is time to overcome the elasticity of the tendon (series elastic components) so that full tension can be developed. Another characteristic of the muscle response is the length–tension relationship. If a muscle is passively stretched the tension increases because of the resistance of the series elastic elements. If the muscle is stimulated to contract, then the active tension developed increases with increasing length, up to a maximum seen when the muscle length is a little greater than the resting length. Thereafter the tension falls away as the muscle length increases. The dependence of tension production on the muscle length is related to the amount of interaction and hence the overlap between thick and thin filaments. When there is no overlap there is no interaction between the filaments and no tension develops. At the very short lengths the thin filaments begin to overlap, interfering with the formation of cross-bridges. TFFTT

Systems of the Body

MCQ 34

The sinoatrial node

1. is situated at the right atrium, close to the coronary sinus.
2. is innervated by the left vagus, whereas the A-V node is innervated by the right.
3. if cooled or crushed leads to bradycardia.
4. spontaneously discharges more frequently than the Purkinje fibres.
5. is depressed by digitalis.

The sinoatrial (S-A) node is a half-moon-shaped region situated near the right auricular appendage and is innervated by the right vagus, the left vagus innervating the A-V node. Of the spontaneously firing cells of the heart those of the S-A node generate electrical impulses at the fastest rate and hence set the pace of contraction of the heart. The A-V node and Purkinje fibres are also sites of spontaneous electrical activity and hence are potential pacemaker cells. The A-V node fires more rapidly than the Purkinje fibres. Apart from the firing rates there are other pieces of evidence for the role of the S-A node as the cardiac pacemaker. Cooling or crushing the node results in bradycardia; application of drugs to the S-A node will alter the heart rate and the S-A node is the first part of the heart to show electrical activity. The rate of firing of the S-A node can be reduced by stimulation of the vagus nerve, an effect also produced by digitalis. Once initiated at the S-A node the cardiac impulse spreads through the right atrium, along ordinary myocardial fibres to the right atrium via the anterior atrial myocardial bundle (or Bachmann's bundle). The impulses pass to the A-V node via the anterior, middle and posterior internodal pathways. There is a specialized conducting system throughout the ventricles, the upper part being the bundle of His and the lower part a complex network of fibres, the Purkinje fibres. The rapid conduction of 1.5–40 m s^{-1} along these fibres and the diffuse distribution mean that there is a more or less simultaneous depolarization of right and left ventricle cells, so that there is a single co-ordinated contraction. FFTTT

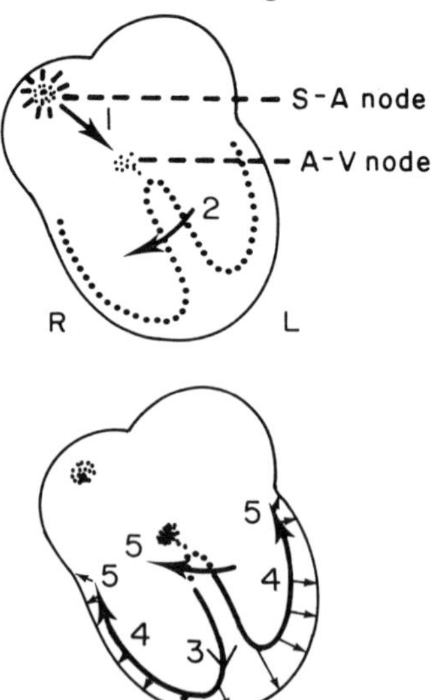

Figure The normal spread of electrical activity in the heart (after Goldman). The numbers indicate the sequence of events.

Heart block

1. is a failure of transmission of impulses in the conducting system.
2. may be caused by vagal stimulation.
3. can result from ischaemia of the atrioventricular junctional fibres.
4. can result from extrasystoles which block normal beat because of the refractory period.
5. can result from compression of the A-V node by scarring.

The atrioventricular (A-V) node is the conducting link between the atria and the ventricles. There is a delay in the transmission of electrical activity from the atria through the node to the ventricles, which allows time for optimal ventricular filling. The relative refractory period of the cells of the mid-portion of the A-V node is very long, protecting the ventricles from excessive contraction frequencies which could occur if the atria were depolarized at high frequencies. The autonomic nervous system exerts its effect on the A-V node. Heart block may occur in the S-A node, the A-V node or the A-V bundle and its branches. Heart block results from impaired excitation conduction and leads to arrhythmia or irregular heartbeat sequence. Conduction may be delayed (first-degree block) or there may be occasional failure of excitation transmission at the S-A or A-V node (incomplete or second-degree heart block) or complete interruption of transmission (complete or third-degree heart block). An incomplete block in the S-A node causes an interruption in the rhythmic discharge of the pacemaker for at least one cycle, whereas in complete S-A block the A-V node generally acts as a second pacemaker. A block in the A-V node results in a marked delay in stimulation of ventricular contraction, which is seen as a prolonged P-R interval (i.e. greater than 0.2 s) in the ECG (see MCQ 37). In complete block, of course, no impulses pass through to the ventricles, which beat at their own slow pace of 30–40 beats min^{-1}, so that the oxygen supply to the body is inadequate for exercise. In this case an electrical stimulator or pacemaker which generates electrical impulses at a preset number of beats min^{-1} can be surgically implanted. If A-V transmission is suddenly interrupted the delay in the ventricular pacemaker becoming active may result in the patient fainting. Heart block may result from vagal stimulation, ischaemia of atrioventricular junctional fibres, or compression of the A-V node by scarring. It can also be produced by hyper- or hypocalcaemia or by high doses of drugs such as the cardiac glycosides or quinidine. TTTFT

MCQ 36

In recording an ECG using bipolar limb leads

1. voltages for lead I are recorded between the right arm and the left leg.
2. voltages for lead II are recorded between the left arm and the left leg.
3. lead II normally corresponds to the cardiac axis and gives the largest ECG deflections.
4. the ECG voltages are of the order of 10 mV.
5. one lead is a reference electrode with an almost constant potential.

The ECG represents the algebraic sum of the fluctuating action potentials of the myocardial fibres during the cardiac cycle. The potentials may be recorded at different loci on the surface of the body as the currents flow out from the heart throughout the body fluids. Einthoven devised the original electrocardiographic lead system for recording the patterns of potentials produced. The vector sum of the electrical activity of the heart at any time was said to lie at the centre of an equilateral triangle bounded by the shoulders and the pubic region. The electrical activity of the heart may be recorded by measuring the potential between pairs of electrodes placed at positions equivalent to the angles of the Einthoven triangle, the arms being used instead of the shoulders and the left leg representing the pubic region. The standard leads are:

I left arm and right arm.
II right arm and left leg.
III left arm and left leg.

The ECG voltages recorded by these electrodes are of the order of 1 mV and are amplified so that a deflection of 1 cm is equivalent to 1 mV. Lead II, which records the potential in a direction 60° to the horizontal, approximates to the normal cardiac axis of the heart and so this lead gives the largest ECG deflections. In contrast to the bipolar lead system, when the voltage is registered between two equivalent points of the body surface, is unipolar recording, when the voltage is registered between an electrode on the body surface and a reference electrode of almost constant potential obtained by joining the other registration sites via resistances. Unipolar limb leads give the potential between one limb and the mean potential in the other two. If the single limb is the right arm, it is termed aVR, the left arm aVL and the left leg aVF. These recording systems give information about the electrical axis of the heart in the frontal plane. The activity in the horizontal plane may be studied employing unipolar chest (pericardial) leads. The single chest electrode is applied to one of six positions across the chest. The indifferent electrode is formed by connecting the three extremity leads through a 5000-ohm resistance to a single terminal (the central terminal of Wilson). TFTFF

MCQ 37

In the normal ECG of the adult

1. the P wave occurs at the beginning of atrial contraction.
2. the P-R interval is within the range 0.22–0.26 s.
3. the Q wave occurs during depolarization of the ventricular muscle.
4. the QRS complex has the same duration as ventricular systole.
5. the Q-T interval increases as heart rate increases.

The electrical activity of the heart may be recorded experimentally using intracellular microelectrodes or by recording from the body surface with extracellular electrodes. Surface ECG recording represents the sum of the action potentials in the individual fibres, but does not resemble the single-fibre action potential. The resting intracellular potential of heart cells is −60 to −90 mV, changing to +20 to +40 mV during activity, so that the surface of excited muscle fibre is electrically negative relative to the unexcited fibre. Thus a potential difference exists across the whole front of an excitation wave spreading over the heart. Each local potential difference forms a dipole which can be characterized in terms of its size and direction by a vector, pointing from the excited to the unexcited state, i.e. from negative to positive. These vectors may be summed to give an integral vector. The voltage registered by the recording electrodes depends on the relative direction of the vector. If the recording is made parallel to the vector the deflection is large, but at right angles there is none. The P wave represents the spread of excitation over the atria. Then follows the isoelectric P-R segment, which is recorded as a straight line and represents the delay in transmission at the A-V junction and the spread of activity in the bundle of His and the Purkinje system. The duration of the P wave plus the P-R segment is less than 0.2 s and represents the P-R interval. The QRS complex, about 0.1 s, represents the spread of the impulse over the ventricle and is thus of short duration compared to the ventricular contraction, which lasts about 0.3 s. The process of repolarization gives rise to the T wave, atrial repolarization being lost in the QRS complex. The R-R interval decreases on exercise. TFFFF

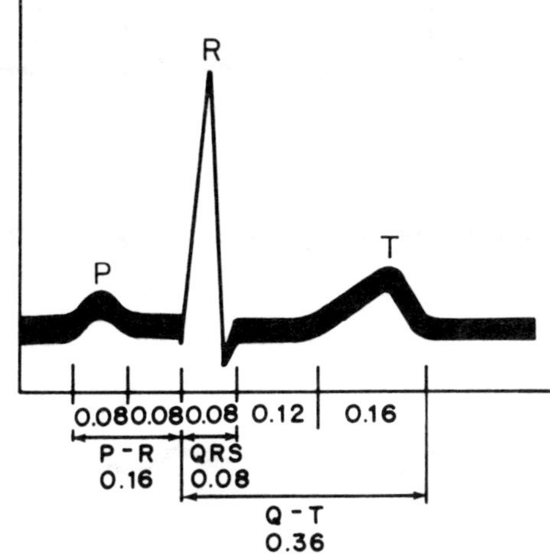

Figure Normal ECG showing duration of waves and intervals.

MCQ 38

Recording the ECG allows determination of

1. disturbances in calcium and potassium ion balance.
2. disorders of coronary blood flow.
3. disorders of valve function.
4. disturbances in repolarization.
5. cardiac volume.

The interpretation of ECG recordings allows the detection of disorders associated with disturbances of the excitation process. It does not, however, provide any information as to the mechanical properties of the heart action. Thus disturbances of excitation generation, conduction and repolarization can be distinguished by recording an ECG. As indicated in MCQ 35, heart block is associated with an increased P-R interval. If the ventricular beat originates at some ectopic focus, then an extrasystole is seen, characterized by a markedly deformed and prolonged QRS complex. An atrial extrasystole is characterized by an unusual P wave. In atrial fibrillation a totally irregular ventricular rhythm is seen. Disorders of coronary blood flow and oxygen blood supply, myocardial infarction, inflammation and any other condition which influences the excitation process will result in abnormal ECGs. This is also true of extracardiac changes affecting excitation processes such as disturbances in autonomic function, of hormonal concentrations, of calcium, potassium and drugs. Disorders of valve function are manifest as heart murmurs, which can be detected by auscultation or with a microphone applied to the chest. Hypertrophy of atria or ventricles can be established through an abnormal ECG, but cardiac volume is determined by X-ray techniques or echocardiography. TTFTF

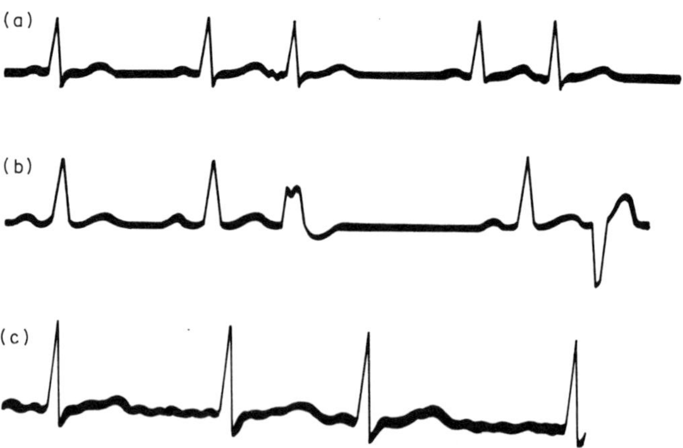

Figure Abnormal ECGs showing extrasystole of (a) atrium and (b) ventricle. Atrial fibrillation is shown in (c).

MCQ 39

The aortic valve

1. comprises three semilunar cusps.
2. has chordae tendineae which prevent it prolapsing into the left ventricle.
3. when open the fibrous flaps are opposed to the walls of the aorta.
4. when incompetent a high pulse pressure results.
5. when incompetent a low systolic pressure results.

The heart resembles an inverted blunt cone lying in the thorax with the base, where the blood vessels enter, pointing upwards and the apex pointing obliquely down. The heart effectively comprises two pumping systems, the right heart supplying the low-pressure pulmonary circulation and the left heart the high-pressure systemic circulation. The blood returning from the the systemic circulation enters the heart at the right atrium via the superior and inferior vena cava. It then passes to the right ventricle and thence via the pulmonary trunk to the lungs. On returning from the lungs the blood enters the left atrium through the pulmonary veins and then passes into the left ventricle and on to the aorta. At the entry and exit of the ventricles are four valves which lie in the fibrous ring separating the atria from the ventricles. The atrioventricular (A-V) valves comprise, in the left heart, the mitral valve, with two cusps or flaps and, in the right heart, the tricuspid valve, which as the name implies has three cusps. They have chordae tendineae which connect the cusps to the papillary muscles and prevent them being forced back into the atria when the ventricular pressure increases. The aortic and pulmonary valves have three semilunar cusps of very thin fibrous tissue, the edges of which are strengthened. It is the action of these valves which transforms the heart into a unidirectional pump. Opening and closing of the valves is brought about passively by pressure changes in the heart. During ventricular systole the A-V valves are closed and the pulmonary and aortic valves open. In diastole the A-V valves are open and the pulmonary and aortic closed. When blood flows through the opening of the valves the flaps are not opposed to the vessel walls or the heart, but are in a midway position. With inadequate closure of the aortic valve or incompetence, the blood regurgitates back, leading to a high pulse pressure, not a low systolic pressure. TFFTF

MCQ 40

During the cardiac cycle at rest

1. the atrial pressure is −8 mmHg (−1.1 kPa) on inspiration.
2. the peak pressure in the right ventricle is between 40 and 50 mmHg.
3. the lowest pressure in the right ventricle is closer to zero than to 20 mmHg.
4. during diastole the left ventricular pressure falls to 25 mmHg.
5. the V wave of a venous pressure tracing corresponds to the peak in atrial pressure.

Blood returning through the systemic and pulmonary circulation enters the heart through the atrium, which acts as a reservoir for blood during ventricular systole. It also helps propel blood into the ventricles, the atrial contraction lasting 0.1 s and generating a pressure of a few millimetres Hg. The atria are thin-walled so the pressure is very similar to that in the surrounding cavity, being −2 mmHg (−0.3 kPa) and falling to −8 mmHg (−1.1 kPa) on inspiration. The atrial pressure is also influenced by ventricular contraction, so pressure increases as the mitral and tricuspid valves bulge into the atria, but falls as soon as blood leaves the ventricles. The veins close to the heart have similar pressure changes to those of the atria. A tracing of the pulse in the jugular vein shows three peaks: the A wave, corresponding to atrial systole; the C wave, being synchronous with the pulse in the carotid artery; and the V wave, corresponding to the peak arterial pressure before the A-V valves open during diastole. At the beginning of ventricular systole the valves close and the ventricles contract isometrically, with the pressure rising rapidly to around 80 mmHg (10.7 kPa) in the left ventricle and to the much lower value of 8 mmHg (1.1 kPa) in the right ventricle. At this point the pulmonary and aortic valves open and the ventricles contract isotonically, the maximum pressure achieved being 120 mmHg (16.0 kPa) and 25 mmHg (3.3 kPa) in the left and right ventricles, respectively. The rate of ejection of blood then falls so that the ventricular pressure too falls to below that in the aorta and pulmonary artery, leading to closure of the valves at the entry to these vessels. Then follows the isometric relaxation period when the pressure falls very rapidly and finally there is the filling period when the ventricle is passively stretched and the pressure represents that in the mediastinum. Essentially the pattern of pressure changes in the two ventricles is similar, but the pressure achieved in the right side of the heart is much lower than in the left. TFTFT

The two heart sounds

1. are due in part to the turbulence produced by valve closure.
2. may be split owing to asynchronous valve closure.
3. are of similar duration.
4. have identical frequency.
5. do not differ from heart murmurs.

During the cardiac cycle vibrations in the range 15–400 Hz are produced which may be heard using a stethoscope or phonocardiograph—a microphone linked to a recording system. Those associated with normal cardiac function are called heart sounds and those with pathological conditions heart murmurs. The two main heart sounds are the first ('lub') at the onset of systole and the second ('dup') at the onset of diastole. The first heart sound is the slightly longer of the two and has a frequency of about 35 Hz. It is largely due to the vibration of the ventricles and valves associated with the onset of systole and closure of the A-V valves. The second, sharper heart sound has a frequency of about 50 Hz and is coincident with the end of the T wave. It is caused by the cusps of the semilunar valves striking and is sometimes split into a first component accompanying the closure of the aortic valve and a second associated with the closure of the pulmonary valve. Third and fourth heart sounds can be heard with a microphone. The third sound, usually only detectable by auscultation in children, is caused by a rush of blood into the ventricle early in diastole. The fourth sound associated with atrial contraction occurs at the end of the P wave. Murmurs are relatively prolonged vibrations that originate in the heart or great vessels as a result of turbulent blood flow. Normal flow of blood in the cardiovascular system is laminar and silent. However, when regurgitation occurs, as a result of inadequate closure of the valves, or blood flows rapidly through a narrowed valve orifice (stenosis), turbulent, noisy flow occurs. Diagnosis may be based on the timing of the murmurs and the character of the sounds. Murmurs occurring during systole result from incompetent mitral or tricuspid valves or from aortic or pulmonary stenosis. Diastolic murmurs are produced by incompetent aortic and pulmonary valves or stenosis of mitral or tricuspid valves. TTFFF

MCQ 42

Cardiac output

1. is approximately 4.5 l in a pregnant woman.
2. can increase five times on exercise.
3. will increase with an increase in aortic pressure.
4. is decreased on exposure to positive gravity (2–6 g).
5. is always increased in response to increased heart rate.

Cardiac output is given by the product of stroke volume and heart rate. At rest in a young adult the stroke volume is about 70 ml and the heart rate is approximately 70 beats min^{-1}, giving a cardiac output of about 5 l. The methods for determining cardiac output are discussed in MCQ 6 and the factors influencing heart rate in MCQ 43. Increased heart rate does not necessarily result in increased cardiac output. Indeed, increased heart rate reduces filling time and diastolic volume so cardiac output can fall. It is dependent on many factors, including age, changes in posture and exercise. It also increases from about 4.5 l to 6 l during pregnancy to meet the various increased needs of the body. Starling's observations revealed that the heart acts as a constant-flow rather than a constant-pressure pump. Therefore changes in pressure in the aorta are unlikely to influence cardiac output, although it is possible that large changes reduce output. A key factor in influencing cardiac output is venous return. Increased venous return results in end diastolic volume and increased cardiac output. The veins contain about 60 per cent of the blood in the circulation, so that the state of the veins influences venous return. If venous tone is increased then venous return may be markedly elevated. Assuming an upright posture leads to pooling of blood in the distensible veins and so cardiac output falls. The effect is even more pronounced on subjecting someone to acceleration several times that of gravity. If special precautions are not taken, such as the use of anti-gravity suits and positioning astronauts at right angles to the direction of travel, then cardiac output falls dramatically and blackout results. Contraction of skeletal muscles assists in the return of blood to the heart and the presence of valves in the veins prevents backflow. Negative intrathoracic pressure produced on inspiration also gives a greater pressure gradient down which the blood can return to the heart. Even ventricular suction can play a part. All these factors contribute to an increase in cardiac output of up to five times during exercise, as well as sympathetic stimulation. FTFTF

MCQ 43

Heart rate

1. is increased by a reduction in the length of systole relative to diastole.
2. can vary over the respiratory cycle.
3. would be decreased in the resting subject after cardiac denervation.
4. increases on assumption of the upright position.
5. increases on sympathetic stimulation as a result of an increase in the rate of diastolic depolarization of pacemaker cells.

The resting potential of the atria and ventricles is a stable one of approximately −80 mV. The S-A and A-V nodes have a potential never falling below −60 mV, while the Purkinje fibres have a potential of −70 to −90 mV. The action potential of the myocardium is prolonged, of similar duration to the time of contraction. Initially there is a rapid rise in the potential produced by an influx of sodium ions, then follows a plateau resulting mainly from a slow influx of calcium ions balanced by an efflux of potassium ions. The shape of the action potential of the pacemaker cells is a little different and in addition there is no stable resting potential. Immediately after an action potential slow repolarization commences, largely as a result of a decrease in potassium permeability in the presence of relatively high sodium permeability. When the threshold potential is reached, a further action potential occurs: the greater the rate of diastolic depolarization, the greater the heart rate. Sympathetic stimulation increases the rate of diastolic repolarization, while parasympathetic stimulation has the reverse effect. The nerves play a part in cardiovascular reflexes, although the fact that heart transplants are successful demonstrates that they are not paramount. Under resting conditions the slowing effect of the vagus (parasympathetic tone) is dominant, so that section of the nerves would lead to an increased heart rate. On assumption of the upright position there is effectively a decrease in the circulating blood volume because of the pooling of blood in the highly distensible veins. In order to maintain blood pressure the heart rate is increased. Increases are also seen on exercise, the change in rate being brought about largely by a shortening of diastole (0.5 s at rest), which can fall to nearly one-quarter of the resting value. Systole (0.3 s at rest) can shorten, but relatively less. In the young respiration produces a fluctuation in vagal tone and hence heart rate, which increases during inspiration. This so-called sinus arrhythmia is said to result largely from irradiation from the inspiratory to the cardiac centre. Heart rate is also influenced by temperature, hormones such as thyroxine, painful stimuli and blood gas tensions. FTFTT

MCQ 44

Cardiac contractility

1. is synonymous with force of contraction.
2. is enhanced by application of acetylcholine.
3. is reduced via stimulation of the sympathetic nerves.
4. is enhanced by elevation of extracellular Ca^{2+}.
5. is reduced by administration of cardiac glycosides.

The term contractility is not synonymous with force of contraction, since the former is specifically defined as an increased force of contraction not due to alteration in the volume of the ventricle at the end of diastole. The relationship, however, between the end diastolic volume or the degree of stretching of the ventricular muscle and the force of contraction and stroke volume was demonstrated by Starling using an isolated heart–lung preparation in which blood is passed to and from the heart in tubes and blood supply to the heart is from a reservoir. By raising this reservoir the pressure producing ventricular filling is increased and hence the ventricle is stretched. Like other muscle when stretched, it responds by increasing its force of contraction and this leads to increased stroke volume. Thus the increased venous filling or cardiac return is balanced by an increased stroke volume. This is important in balancing the output of the left and right side of the heart. The myocardial cells have a sympathetic nerve supply, stimulation of which alters the relationship between the stroke volume and end diastolic volume, so that the stroke volume is greater for any given end diastolic volume. Thus stimulation of the sympathetic nerves not only results in a more forceful contraction, but also produces more complete ejection. Acetylcholine would tend to reduce contractility, but since few fibres supply the ventricular muscle parasympathetic stimulation has little effect. Any influence on contractility is called an inotropic effect, so the sympathetic nervous system is said to have a positive inotropic effect. Calcium, which is essential for the coupling of excitation to contraction in muscle, has a similar effect, as does the administration of cardiac glycosides. FTFTF

MCQ 45

Pathological conditions which can reduce cardiac output include

1. patent ductus arteriosus.
2. coronary heart disease.
3. anaemia.
4. hypertrophy of the heart.
5. valvular heart disease.

Any defects which affect the ability to pump blood naturally affect cardiac output. Conditions influencing peripheral resistance, blood volume and blood viscosity also affect cardiac output. Congenital heart defects are found in 0.5 per cent of babies who survive the first month of life. They result from incomplete separation of the two sides of the heart at birth or from anomalies of the major vessels. In the fetus the blood is oxygenated in the placenta and so bypasses the fetal lung. It passes from the right to the left atrium via the foramen ovale and also from the pulmonary artery directly into the aorta through the ductus arteriosus. Sometimes these shunts fail to close at birth, leading to a reduced cardiac output. Valvular heart disease with the associated regurgitation of blood places an additional burden on the heart and reduces cardiac output. Among the most common causes of acquired valvular defects are acute rheumatic fever and bacterial endocarditis. Coronary or ischaemic heart disease results from narrowing of coronary vessels or additionally from thrombosis, leading to ischaemia of the myocardium and reduced cardiac output. It may range in severity from a mild condition, sometimes accompanied by a pain in the chest or angina pectoris, to an acute coronary condition, myocardial infarction and possibly sudden death. An increase in the size of the heart fibres is called cardiac hypertrophy. It represents one of the main ways in which the heart adapts to an increasing work-load. It is commonly seen in athletes, but also occurs with the pathological changes just described. In these cases the tissue is under-vascularized so that the heart cells are deprived of oxygen, fibrous tissue forms and the heart begins to fail. Anaemia resulting from a reduction in the number of cells or of the haemoglobin content is associated with reduced blood viscosity and hence a fall in peripheral resistance and accompanying increased cardiac output. TTFTT

MCQ 46

With regard to the energetics of heart action

1. glucose and free fatty acids are important substrates at rest.
2. lactic acid is the main substrate during exercise.
3. during diastole the ventricular muscle develops mainly kinetic energy.
4. the kinetic energy decreases with age.
5. the heart normally extracts 70–90 per cent of oxygen from arterial blood.

Cardiac muscle depends on aerobic metabolism. The supply of oxygen can be improved by increased blood flow, increased myocardial efficiency and increased oxygen extraction, so that coronary artery venous oxygen concentration is 0.05 l as opposed to 0.08 per litre of blood at rest. The venous reserve of oxygen is therefore very low, with 70–90 per cent of oxygen being extracted. Normally 35 per cent of the energy requirements of the heart is provided by free fatty acids that enter cells across the sarcolemma and a further 30 per cent from glucose. Lactate is also metabolized at rest, the proportion increasing with an increased work-load. The heart can also use lactic acid, breakdown being via the Krebs cycle. Hence if insufficient oxygen is available for this to occur, there are higher concentrations of lactic acid found in the vein than in the artery. During systole the ventricular muscle carries out pressure–volume work, tension being developed and the fibres shortening. It also develops kinetic energy to bring the blood to the velocity necessary for ejection. With the loss of elasticity of the aorta in old age acceleration work increases, as it does in exercise. TTFFT

MCQ 47

The resistance to flow in a blood vessel

1. depends on the quantity of smooth muscle in the vessel wall.
2. is inversely proportional to the square of the radius.
3. is directly proportional to the vessel length.
4. is dependent on the haematocrit.
5. is directly proportional to the pressure drop along the vessel.

As with electrical current, for which the resistance is described as the voltage drop divided by the current flow, the resistance to flow of liquids is given by P/Q, where P is the pressure gradient across the tube and Q is the volume flow rate. The factors contributing to resistance to flow through cylindrical tubes were studied by the French physician Poiseuille, who stated that it is directly proportional to the length (l) and viscosity (η) and inversely proportional to the fourth power of the radius. Poiseuille's equation may be rewritten to give

$$\frac{P}{Q} = R = \frac{8\eta l}{\pi r^4}$$

Thus changes in the diameter of the vessel have a marked influence on resistance, since the term for radius is raised to the fourth power. Since vessel length does not alter greatly, changes in resistance are largely due to changes in diameter and blood viscosity. Normally there is little change in viscosity. This term was described by Newton as an internal friction of liquids causing a defect of slipperiness between adjacent layers or laminae of a moving fluid. Viscosity may increase in the pathological conditions of leukaemia and polycythaemia, resulting in an increased resistance. Viscosity is also influenced by the bore of vessels, decreasing as the vessel diameter falls below 150 μm. This, however, may be balanced by the increase in blood viscosity resulting from the decrease in flow rate. Normally blood flow in vessels is laminar with the laminae moving parallel to each other, the higher velocity of flow being in the centre of the vessel. If, however, the flow velocity is very high or the blood passes an obstruction turbulence occurs, which considerably increases the resistance. FFTTT

MCQ 48

Mean arterial pressure

1. is equal to the product of cardiac output and total peripheral resistance.
2. is given by the formula diastolic pressure +⅓ pulse pressure.
3. decreases with increasing age.
4. is higher in women than in men under 40–50 years of age.
5. is unrelated to the weight of the subject.

The equation for flow in tubes is P = flow × resistance, where P is the pressure difference. As the vascular tree is a continuous series of closed tubes, the equation holds for the entire tree. Flow is effectively the cardiac output, the resistance is the total peripheral resistance and the pressure difference is the aortic pressure minus the pressure at the end of the vena cava. Since pressure in the vena cava is 0 mmHg, P is the mean arterial pressure. The area under the blood pressure curve is less than one would expect so the mean arterial pressure is not ½ (diastolic + systolic pressure), but diastolic pressure + ⅓ pulse pressure. Blood pressure varies with many factors, including age, weight, sex, race and socio-economic status. Blood pressure normally increases from about 115/70 mmHg (14/9.3 kPa) at the age of 15 to 140/90 mmHg (18.7/12.0 kPa) at 65 years of age. The effect on arterial pressure is largely due to the loss of elasticity of the vessels, and the effect on diastolic pressure to a change in the total peripheral resistance. Up to about 50 years of age women tend to have a lower blood pressure than men; the reverse is true above that age. Blood pressure also appears to be related to the weight of the subject, increasing as the weight of an individual increases. Loss of elasticity on ageing is also manifest in an increase in pulse wave velocity. The pressure changes in an artery cause expansion and contraction of the arterial wall. The pressure change is transmitted from the aorta at about 6 m s^{-1} and hence takes 0.1 s to reach the wrist, whereas the blood takes several seconds. TTFFF

MCQ 49

Measurement of arterial blood pressure in man

1. may be made directly using a transducer.
2. can employ the auscultatory method.
3. can be performed with the palpation method.
4. entails recording of systolic pressure when the sounds in the brachial artery become muffled.
5. yields inaccurately high values if too wide a cuff is used.

Measurement of arterial blood pressure is important both clinically and experimentally. It may be measured directly by introducing a catheter into the appropriate vessel and making a recording via a transducer. Because this entails puncturing a vessel this technique is limited to a few clinical situations. The routine method of measurement employs the inflatable cuff and manometer or sphygmomanometer introduced by Riva Rocci. Together with the auscultatory method of Korotkov it enables the measurement of systolic and diastolic pressure. The cuff is rapidly inflated to a pressure above the expected systolic pressure so that the flow in the brachial artery is obstructed. The pressure in the cuff is slowly reduced. At the moment when the pressure falls to the systolic pressure blood is found to flow through the compressed region and a tapping sound is heard. When the cuff pressure falls to diastolic pressure, flow is no longer impeded and laminar flow is resumed, so that the sounds become muffled and disappear. In the UK the point at which the sounds become muffled is taken as the diastolic pressure; in the USA it is generally taken as the point when the sounds disappear. The palpation method may be used to determine systolic pressure. When the pressure in the cuff reaches systolic pressure and blood flows into the brachial artery a pulse is felt. To obtain accurate measurements the cuff should be at the level of the heart. Cuff width is also important. It should be about half the circumference of the arm. If the cuff is too broad the pressure measured is too low. Sphygmomanometric methods of determining pressure can be automated, but cannot be used to record pressure continuously. TTTFF

MCQ 50

Resistance to blood flow

1. is greatest in the capillaries.
2. in two vessels in parallel is less than the resistance of either vessel alone.
3. prevents hardening of the arteries.
4. helps prevent arterial pressure falling to zero during diastole.
5. may be expressed in peripheral resistance units in mmHg ml^{-1} min^{-1}.

During systole a peak of 120 mmHg (16 kPa) is achieved by the left ventricle. However, on diastole pressure falls away to zero. A more or less steady outflow is achieved by a combination of the elastic recoil of the aorta and the resistance to flow offered by the peripheral arterioles. On leaving the heart the blood enters the 'windkessel' vessels, which have highly elastic walls. Potential energy is stored by the elastic tissue during ventricular contraction and reconverted to kinetic energy to maintain circulation during diastole. Thus the pressure head is steadier than it would otherwise be. Pressure in these large arteries is maintained at approximately 120/80 mmHg (16/10.7 kPa). The greatest resistance to flow, when the major drop in arterial pressure occurs, lies in the arterioles, which are the precapillary resistance vessels. The pressure as they enter the capillaries is approximately 30 mmHg (4 kPa) and the pulsatile flow has been lost. Changes in the diameter of the arterioles determine the blood supply of the different regional circuits. Fine adjustment of the blood flow to active or inactive tissue can occur as the blood vessels are arranged in parallel. With resistances arranged in parallel, the total resistance (R_{tot}) is given by

$$\frac{1}{R_{tot}} = \frac{1}{R_1} + \frac{1}{R_2} + \frac{1}{R_3} + \frac{1}{R_4} + \cdots$$

where R_1, R_2, etc. are the individual resistances. Thus the total resistance is less than the individual resistances. Resistance is expressed in peripheral resistance units, which give the resistance to flow through 100 g tissue in ml min^{-1} for a given pressure gradient (in mmHg). If, for example, the pressure gradient across the resting muscle is 100 mmHg and the rate of flow is 3 ml 100 g^{-1} min^{-1} then the resistance is 33.3 peripheral resistance units (PRU). In exercise the rate of flow may increase to 70 ml 100 g^{-1} min^{-1} and hence the resistance falls to 1.4 PRU. FFFTT

MCQ 51

Veins

1. have a greater wall thickness to lumen diameter than arterioles.
2. may contain up to 75 per cent of total blood volume.
3. are dilated by parasympathetic nerves.
4. have sparse innervation as compared to arterioles.
5. contain smooth muscle in their walls which contracts after venous distension.

The walls of the arteries and veins consist of three coats: an outer layer of connective tissue, the tunica adventitia; a middle layer of elastic and smooth muscle cells, the tunica media; and an inner layer of endothelial cells, the tunica intima. The intima media of the arteries is relatively thick, whereas the veins are characterized by thin walls, which have little elastic tissue and are quite easily distensible. Their wall thickness to lumen diameter is thus less than that of the arterioles. The walls contain a small amount of smooth muscle which contracts when the veins are distended. The muscle is supplied by adrenergic nerves which produce venoconstriction. This response is also produced by noradrenaline and venospasm may be clearly seen, for example, as a result of local injury on sampling of blood. The veins collapse when the transmural pressure falls below 6 mmHg. If the transmural pressure increases the volume of the veins increases markedly. This results not so much from the distensibility of the veins but from a change in the cross-sectional profile from elliptical to circular. This property of the veins means that a large percentage of the total circulating blood volume can be sequestered in them. It also means that small changes in the transmural pressure bring about marked changes in venous capacity. Thus, on assumption of the upright posture the volume of the blood in the veins increases by some 400 ml as a result of the increased hydrostatic and hence transmural pressure in the foot. However, the thin walls also mean that vessels are easily compressed by the surrounding skeletal muscle when it contracts and this aids in pumping blood back to the heart. FTFTT

MCQ 52

Capillaries

1. directly join arterioles to venules.
2. have a uniform distribution through the body.
3. in the brain have thin, non-fenestrated endothelium with a continuous basement membrane.
4. in the heart have discontinuous endothelium.
5. in the kidney have fenestrated endothelium.

The capillaries are the exchange vessels. They do not arise directly from the arterioles, but frequently branch from the so-called preferential channels, the metarterioles. At the junction with the capillaries are the precapillary sphincters formed of smooth muscle. It is the degree of contraction of these sphincters which determines the amount of blood going through the capillaries. Blood does not necessarily go through the capillaries, but may go through the arteriovenous anastomoses. The supply of blood to a tissue depends on the capillary density. The myocardium, brain, liver and kidneys have high capillary density, whereas bone, fat and connective tissue have a relatively low density. The walls of the capillaries are made up of a single layer of epithelial cells, but the structure varies from organ to organ. The epithelial cells may form a continuous layer, except at the intercellular region. Some of these intercellular spaces have channels some 4 mm wide through which fluid may pass. Fenestrations are found in the walls of some capillaries which are intracellular openings through which it seems transudation of fluid occurs. Such capillaries occur in the renal glomeruli and in endocrine and exocrine glands. Finally there are continuous capillaries or sinusoids, which occur in the marrow, liver and spleen. In these vessels there are large gaps between the individual cells which allow the exchange not only of large protein molecules, but also of the red blood cells themselves. FFTFT

MCQ 53

In the triple response the flare is produced by

1. a local monosynaptic reflex.
2. capillary engorgement due to venous contraction.
3. a direct capillary response to local damage.
4. local formation of angiotensin II.
5. an axon reflex affecting arterioles.

If a pointed object is drawn lightly over the skin, after about 15 s a white line appears, corresponding exactly to the area stimulated (white reaction). The response can still be seen after nerves supplying the skin have been sectioned, so this is a local response. The fact that the response is very localized and can be obtained after the arterial flow has been obstructed indicates the response is due to capillary and not arteriolar contraction. It seems that the stimulus initiates contraction of the precapillary sphincters so that blood drains out of the capillaries and small veins. When the skin is stroked more firmly the response is very different. First a red line instead of a white line appears, which seems to be a purely passive response dependent on the relaxation of precapillary sphincters, as it is not blocked by local anaesthetics. Then, after a delay, a redness or flare spreads out from the injury, the area corresponding to arteriolar distribution. The response appears to be mediated by nerves as it is abolished by local anaesthetics. It is, however, unaffected if the relevant nerve trunk is blocked, showing that the response does not entail central nervous connections. The response is, instead, due to an axon reflex. According to the concept of an axon reflex sensory nerve fibres in the skin may have collateral fibres passing to the adjacent blood vessels. Impulses in addition to passing to the spinal cord may pass directly to these blood vessels, causing vasodilatation. There is evidence that the antidromic responses release substance P at the endings near the cutaneous arterioles. After the red line and the flare the third response seen in this triple response to injury is a wheal or blister which develops through leaking of fluid from the capillaries. The triple response may be simulated by injection of histamine under the skin. FFFFT

MCQ 54

Nervous impulses from the arterial baroreceptors

1. arise in the carotid and aortic bodies.
2. may travel in the vagus nerve.
3. increase in frequency as the arterial carbon dioxide tension increases.
4. are seen at a threshold blood pressure, then increase in frequency as pressure increases.
5. influence the rate of cardiac contractions.

Baroreceptors are mechanoreceptors within the arterial tree, situated in the adventitia and sensitive to the stretch normally caused by a rise of pressure expanding the arterial wall. They are found in the carotid sinus, the aortic arch, the root of the right subclavian artery and in the junction of the thyroid artery with the common carotid artery. Chemoreceptors sensitive to changes in the arterial gas tensions lie in the carotid bodies. Nerve impulses from the carotid sinus travel in the vagus nerve. Afferent fibres from other baroreceptor regions travel in the left and right aortic nerves which lie in the vagus sheath. The impulses pass to the so-called medullary vasomotor centre (an over-simplified term) and to the nucleus ambiguus, which is the cardiac vagal centre. Impulses from the baroreceptors inhibit the vasomotor centre. Thus when blood pressure increases arteriolar vasoconstrictor activity is reduced, so that the change in blood pressure is minimized. There are effects on heart rate which also help to keep the blood pressure constant, an increase in blood pressure resulting in a slowing of the heart. The efferent pathways of these reflexes are the parasympathetic pathways to the heart and the sympathetic fibres to the heart and blood vessels. A mean arterial blood pressure of 60–70 mmHg is required before sufficient impulses travel from the carotid sinus to the medulla to inhibit the cardiovascular centres. This is sometimes called the threshold of the sinus reflex. Above this point the impulses increase in frequency as the pressure increases. In man the main baroreceptors lie in the internal carotid sinus and are thus able to monitor the blood supply to the brain. These baroreceptors may be artificially stimulated by pressure applied to the neck and in susceptible subjects the resulting blood pressure drop may lead to unconsciousness. It is said that Victorian ladies twisted their pearl necklaces to produce such an increase in pressure and thereby facilitate the socially desirable fainting fit. FTFTT

MCQ 55

The areas of the central nervous system involved in integration of cardiovascular control include the

1. spinal cord.
2. medulla.
3. hypothalamus.
4. pons.
5. premotor cortex.

The short-term control of blood pressure is via the integration of a number of cardiovascular reflexes. Afferent inputs travel from the baroreceptors (see MCQ 54) to the brain and from the cardiac mechanoreceptors located in the walls of the aorta and venae cavae. These volume receptors provide information on the state of filling of the vascular system and the dynamics of the filling of the ventricles. Inputs from the carotid sinus and aortic bodies also contribute to the regulation of blood pressure so that on asphyxia or anoxia a reflex sympathetic response is seen. There are also a number of non-specific afferents, including those from the thermoreceptors, nociceptors and the respiratory centre. The efferent pathways form part of the sympathetic autonomic nervous system. Preganglionic fibres leave the spinal cord in roots T_1 to L_2 and enter mixed peripheral nerves to supply the smooth muscle of the vascular system. The nerves to the heart arise in T_2 to T_4, synapsing in the cervical ganglion and reaching the heart via the cardiac branches of the sympathetic chain. The parasympathetic supply to the heart travels via the vagus. The primary area for the integration of reflexes is the so-called vasomotor centre, which comprises diffuse groups of cells in the lower third of the pons and upper part of the medulla (bulbopontine area). For convenience it may be divided into pressor and depressor areas. The pressor area exerts predominantly an excitatory effect on sympathetic vasoconstrictor activity, but also contains neurones which influence the heart. Anatomically there is no clear separation between the pressor and depressor areas. The cardiac centre (cardio-inhibitory centre) is the nucleus ambiguus of the vagus, lying lateral to the medullary reticular neurones which modify the spinal sympathetic discharge. Although the circulation is chiefly regulated by centres in the brain stem a degree of circulatory control is possible in the spinal animal and influence is exerted by other regions, particularly the hypothalamus. The hypothalamus adjusts the cardiovascular system to the prevailing autonomic response, e.g. fight or flight. The control function of the cortex becomes evident in the anticipatory changes in the cardiovascular system seen, for example, before the start of a race. FTTTF

MCQ 56

Vasodilatation can be produced by

1. noradrenaline.
2. histamine.
3. a reduction in sympathetic nervous activity.
4. a low glucose concentration.
5. elevated CO_2 concentrations.

Adjustments of flow can be locally regulated so that an individual organ receives sufficient blood to meet its metabolic requirements. The responses are also integrated so that the various organs receive a blood flow commensurate with their relative activity, while blood pressure is maintained. In some organs the control of blood flow is largely under the control of local factors, of substances involved in cell metabolism. Thus vasodilatation can be produced by an increase in P_{CO_2}, a fall in the local partial pressure of oxygen, a fall in pH or an elevation of the concentration of adenosine, adenosine diphosphate (ADP), adenosine monophosphate (AMP) or K^+ ions. Vascular responses may also be produced locally or in the systemic circulation by humoral agents, including the vasoactive peptides, the kinins and angiotensin II, which are formed from plasma proteins under the action of the enzymes kallikrein and renin, respectively. Kallikrein is released from an actively secreting gland and splits off the polypeptide kallidin from an α_2-globulin. Subsequently it is converted to bradykinin, which is a potent vasodilator. Angiotensin II is formed as a result of the action of renin, which when released from the kidney also acts on an α_2-globulin to form a decapeptide, angiotensin I, which through the action of a converting enzyme is broken down to an octapeptide, angiotensin II. Histamine is also a powerful vasodilator and is released largely as a result of damage to skin and of some antibody–antigen reactions. Catecholamines too have a powerful effect on blood vessels, whether released from the adrenal medulla or at sympathetic nerve endings. Smooth muscle of most blood vessels is supplied with sympathetic nerves which release noradrenaline. Stimulation of such nerves leads to vasoconstriction: vasodilatation is produced by a reduced sympathetic outflow. FTTFT

MCQ 57

The pulmonary circulation differs from the systemic circulation in that

1. the arteries serve as important blood reservoirs.
2. the adrenergic sympathetic neurones are more important in controlling arteriolar and precapillary resistance.
3. hypoxia produces vasoconstriction.
4. the capillaries have a higher transmural pressure.
5. obstruction of blood flow is much more likely to cause retrograde ventricular failure.

The volume of blood in the pulmonary circulation at rest is 600 ml, of which nearly as much is in the arteries as in the veins, the pulmonary arteries being very distensible. The pressure in the pulmonary arteries is much lower than in the systemic arteries, the systolic pressure in the pulmonary circulation being 20–25 mmHg (2.7–3.3 kPa) and the diastolic pressure 6–12 mmHg (0.8–1.6 kPa). The pulmonary capillary pressure is about 8 mmHg (1.1 kPa), well below the colloid osmotic pressure of 25 mmHg (3.3 kPa). Since the transmural pressure is below that of the colloid osmotic pressure the pulmonary capillaries are unlikely to lose fluid, which is important as the alveoli do not possess lymphatics. Pulmonary oedema does occur if the pressure in the circuit increases above 24 mmHg (3.2 kPa). In the pulmonary circulation the main point of resistance lies in the capillaries and many of the arterioles are almost devoid of smooth muscle. There is a sympathetic adrenergic nerve supply, stimulation of which decreases the compliance of the arteries and veins. Large changes in pulmonary resistance are possible, as during exercise when the cardiac output is increased five or six times pulmonary pressure changes little, which means resistance has fallen markedly. Hypoxia produces vasoconstriction of pulmonary vessels, which can result in shunting of blood from inactive parenchyma to regions where gaseous exchange occurs. Chronic hypoxia will, however, lead to pulmonary hypertension. Pulmonary vessels are also constricted by histamine. Obstruction of over two-thirds of the pulmonary vascular bed would have to occur before left ventricular failure developed. Left ventricular failure is about 30 times more common than right ventricular failure.
TFTFF

MCQ 58

Left coronary artery blood flow

1. is greater in diastole than systole.
2. is increased by adenosine.
3. is significantly controlled via sympathetic vasodilator nerves.
4. is increased by hypocarbia.
5. is increased by high pH in the coronary circulation.

There are generally two coronary arteries in the human. The right one supplies the greater part of the right ventricle, septum and posterior wall of the left ventricle. The remaining regions are supplied by the left artery. The heart receives a very high blood flow of about 200 ml min^{-1}. The factors affecting coronary blood flow are physical and metabolic. The flow of the coronary circulation shows marked fluctuations. These are partly due to the fluctuations in the aortic pressure, but the main cause of course is the increase in interstitial myocardial pressure during contraction, which compresses the blood vessels in the heart. This is more marked in the more muscular left ventricle, in which only 25 per cent of the total left coronary flow occurs during systole. Because of these influences on coronary flow, diseases which have an effect on coronary arterial pressure, such as aortic stenosis, or an increase in venous pressure as in congestive heart failure, offering impedance to the coronary circulation, have a markedly constraining effect. The blood flow through the circulation is almost exclusively regulated by local intrinsic control of blood vessels in response to metabolic requirements. However, there is an autonomic nerve supply to the vessels and changes in the resistance can be brought about by local release of neurotransmitters, such as catecholamines. The endogenous catecholamines can also affect blood flow by influencing heart rate and force of contraction. The principal regulator of coronary artery flow is the oxygen requirement, oxygen deficiency causing marked dilatation. Dilatation is also brought about by other metabolites, including adenosine. If blood flow to the heart is obstructed, then the accumulating metabolites cause ischaemic pain or angina pectoris, which can be relieved using a vasodilator, nitroglycerine. If the period of anoxia lasts for more than 30 min the myocardium undergoes irreversible structural changes. TTFFF

MCQ 59

Dizziness due inter alia to cerebral vasoconstriction could be produced by

1. hypercarbia.
2. hypoxia.
3. hypertension.
4. decreased impulses in the sympathetic pathway to the cerebral arteries.
5. increased impulses in the parasympathetic pathways to the cerebral arterioles.

The brain receives a relatively good blood supply—about 14 per cent of the total cardiac output at rest. This is important as the neurones are very sensitive even to brief deprivation of oxygen and nutrients. In the human the blood supply to the brain is from the internal carotid arteries and is additionally assured by the collaterals from the circle of Willis. It is necessary that the blood flow remain constant, not only because of the supply of nutrients, but also because the brain is uniquely enclosed within a rigid container and so its volume must not change. Another feature of the cerebral circulation is that most capillaries do not have fenestrations and the plasma membrane of adjoining endothelial cells is partially fused, which is one of the reasons why the blood–brain barrier exists. There is an abundant adrenergic nerve supply to the large cerebral arteries, but vasomotor control of the cerebral circulation is thought to be of little importance. Elevated $P\text{CO}_2$ or depressed $P\text{O}_2$ causes cerebral vasodilatation, but the prime chemical regulator is the $P\text{CO}_2$, acting by means of changes in the extravascular hydrogen ion concentration. Conversely, hypocapnia leads to vasoconstriction. Vigorous hyperventilation will lead to dizziness, partly as a result of O_2 lack and partly due to the acapnia and alkalosis. In addition to the overall effect on blood flow, the local effect of a fall in hydrogen ion concentration probably causes an increase in the tone of the precapillary sphincters, leading to reduced capillary perfusion. Also in the presence of low carbon dioxide tensions the haemoglobin dissociation curve moves to the left, so that there is reduced oxygen release. As long as the $P\text{O}_2$ and $P\text{CO}_2$ remain at normal levels the cerebral blood flow remains constant over a wide range of blood pressure, as a result of vasoconstriction at elevated pressures. FFFFF

MCQ 60

During exercise muscle blood flow increases

1. up to 30 times.
2. mainly through neural mechanisms.
3. in response to cholinergic vasodilator fibres.
4. through adrenergic vasodilator mechanisms.
5. through metabolites, including K^+ ions.

The skeletal muscles form about 40 per cent of the total body weight, but under resting conditions they receive a relatively small proportion of the total cardiac output. However, on exercise the flow rate can increase some 30 times and the cardiac output increases to 25 l min^{-1}. Skeletal muscle very clearly has a dual control of the blood vessels, having both an important nervous supply and being regulated by local factors. The vessels receive sympathetic vasoconstrictor fibres and also sympathetic cholinergic vasodilator fibres. Because of the potential capacity of the muscle vascular bed it is important that it be kept in a state of vasoconstriction, otherwise there would be a profound fall in blood pressure. This supply is also important if it is necessary to reduce peripheral resistance overall in the body. Vasoconstriction is brought about by an effect on the α-vascular receptors. There are, however, β-receptors and these can be stimulated by circulating catecholamines to produce vasodilatation. This may be important at the beginning of exercise, but not necessary for the active hyperaemia, which can entirely be accounted for by local factors. The sympathetic vasodilator fibres act on arterioles, not precapillary sphincters, and thus provide a channel to the venules, the significance of which is probably to prevent too great a rise in blood pressure at the beginning of exercise, when the cardiac output rises rapidly. The mechanisms responsible for the metabolically linked vasodilatation are not fully understood, but it appears that no single chemical factor is involved but rather a number of factors acting in concert, primarily K^+, hyperosmolality and oxygen lack. TFTTT

MCQ 61

Thermoregulatory vasodilatation in the hand

1. is caused by activation of cholinergic pseudomotor fibres.
2. occurs when the hypothalamus is heated.
3. is brought about through the action of bradykinin.
4. is caused by reduced sympathetic adrenergic activity.
5. is produced in the same way as cold vasodilatation.

Blood flow through the skin largely serves to maintain the body temperature constant, as opposed to supplying the metabolic requirements of the skin. Under resting conditions skin blood flow shows considerable regional differences depending on skin temperature. On exposure to heat skin flow may increase to 3–4 l min^{-1} and on cold exposure may fall to as little as 50 ml min^{-1}. The additional load on heat exposure makes considerable demands on the cardiovascular system. However, so important is thermoregulation that skin blood flow is maintained to the point of circulatory failure. The responses involve cutaneous and hypothalamic thermoreceptors. Surprisingly there are no specific vasodilator fibres to the skin. In the distal regions such as the hands and the feet vasodilatation is produced by central inhibition of adrenergic vasoconstrictor activity. The large increase in blood flow which is seen results from the opening of numerous A-V anastomoses. In contrast blood flow in more proximal regions and the trunk is increased through the action of bradykinin, which is released on stimulation of the sympathetic cholinergic fibres to the sweat glands. Skin may be exposed to low temperatures without any adverse effect. However, if skin temperature drops to below 10 °C pain may be experienced and vasodilatation ensues. This is due to axon reflexes acting especially on A-V anastomoses. Tissue damage resulting from inadequate supply of oxygen and nutrients causes liberation of histamine, which excites sensory terminals initiating an axon reflex.
FTFTF

MCQ 62

Autoregulation of blood flow

1. occurs in the kidney.
2. is important in the muscle.
3. is partially made possible by the fact that some metabolites are vasodilators.
4. is in part due to myogenic responses.
5. is due to the enhancement of smooth muscle contraction by hypoxia.

Autoregulation of blood flow represents the ability of individual organs to alter their vascular resistance and thus regulate blood flow independently of external neural and hormonal influences. It is obviously important that blood flow to the brain and the heart is maintained and autoregulation in these organs is evident. It is also noted in the kidney and can be seen in the intestine and liver to overcome potentially harmful vasoconstrictor effects associated with the shunting of blood away from these organs. This latter is called autoregulatory escape. The range of arterial pressure over which blood flow can be kept constant in certain organs through autoregulation is very wide. Renal blood flow and glomerular filtration rate remain remarkably constant over the range 90–200 mmHg (12–26.7 kPa) and if blood gas tensions remain normal cerebral blood flow is constant over the range 65–140 mmHg (8.7–18.7 kPa). There are a number of mechanisms which may explain autoregulation. The most probable mechanism for autoregulation of the kidney is that it results from alterations in the myogenic tone of the afferent arterioles in response to changes in the perfusing pressure. Evidence of a role for arterioles, i.e. preglomerular effects, is provided by the observation that capillary tubular pressure and glomerular filtration rate remain constant over a wide range of pressures. The fact that autoregulation persists after denervation of the kidney but is abolished after treatment with papaverine or sodium cyanide indicates dependence on the intrinsic myogenic tone of the renal vessels. Vessels of the cerebral circulation also exhibit myogenic autoregulation, but the chief mechanism of cerebral autoregulation is believed to be the metabolic action of carbon dioxide and in more serious conditions of reduced perfusion the effect of pyruvate and lactic acids in lowering the pH.
TFTTF

MCQ 63

On assumption of the upright posture reflex cardiovascular responses include

1. a decrease in stroke volume.
2. a decrease in diastolic pressure.
3. an increased heart rate.
4. vasoconstriction in the muscles of the limb.
5. a fall in peripheral resistance.

On standing there is a redistribution of blood, with some 400–500 ml being transferred from the chest to the legs and a resultant fall in cardiac output of about 25 per cent. This would lead to orthostatic hypotension unless compensatory reflexes came into play. These reflex responses are brought about through the mediation of the baroreceptors and volume receptors. The reflex responses include an increase in heart rate and a small decrease in the stroke volume. Despite this the cardiac output remains reduced as does the central venous pressure. The most important of the reflexes are the vasoconstrictor reflexes, which result in a 25 per cent increase in the peripheral resistance. Vasoconstriction occurs in the skeletal muscles, the skin, the kidneys and the splanchnic region. There is also vasoconstriction in the capacitance vessels that act as reservoirs, namely the veins of the skin and splanchnic region. The response illustrates the importance of the nervous system in producing an integrated response, shunting blood from certain organs to maintain the function of the cardiovascular system. Those tissues maintaining blood flow constant are those in which flow is governed by local factors and there is no significant neural control. The responses are reinforced by a higher rate of catecholamine secretion from the adrenal medulla, activation of the renin–angiotensin system and increased secretion of aldosterone and vasopressin. These hormonal responses have some latency. Venous return can also be enhanced by the muscle pump. The postural reflex disappears after space flights and is reduced after several days in bed. TFTTF

MCQ 64

The rate of diffusion of respiratory gases

1. is related to the random movement of the gas molecules.
2. is proportional to their partial pressure.
3. is dependent on the solubility in the alveolar membrane.
4. is proportional to the thickness of the membranes traversed.
5. is proportional to the molecular weight of the gas.

Oxygen and carbon dioxide are transferred across the alveolar and pulmonary capillary membranes by diffusion. The driving force is the concentration gradient and the rate depends also on the molecular size of the substance, the nature of the medium—gas or liquid—and in the latter case its viscosity. The rate of diffusion is not proportional to the molecular weight of the gas, rather it is inversely proportional to the square root of the molecular weight. The basic event in diffusion consists of random movement (Brownian movement) which in turn is dependent on the temperature. The time taken for diffusion is proportional to the square root of the distance over which it occurs. Thus in biological systems it is only an effective force over very short distances. When a dissolved gas diffuses through a fluid layer the effective concentrations are the corresponding partial pressures. The partial pressure of a gas is defined by Dalton's law, according to which each gas in a mixture exerts a partial pressure proportional to its share of the total volume, i.e. its concentration. The rate of diffusion of a gas across the lungs varies according to the diffusing capacity, which is defined as the quantity of gas transferred each minute for each millimetre Hg difference between the partial pressure in the alveoli and that in the capillary blood. It depends not only on the nature of the gas under consideration, but also on the solubility in the alveolar membrane and the ease with which it diffuses through it. Thus carbon dioxide diffuses more rapidly than oxygen, because it is more soluble. Movement of a respiratory gas also depends on the surface area of the lungs and the average thickness of the alveolar membrane. TFTFF

MCQ 65

The functional residual capacity

1. is about 1.2 l in a healthy adult.
2. generally decreases with age.
3. may be calculated as the residual volume plus the inspiratory capacity.
4. is the volume of gas left in the lungs at the resting respiratory level.
5. can be measured with a spirometer.

In the healthy 70-kg adult, the total lung volume is about 6 l. For descriptive purposes this may be divided into volumes, which are components exchanged during respiration, and capacities, which are determined by the size of the lungs and thorax. The vital capacity (4.8 l in men, 3.0 l in women) is the maximum quantity of air that can be expired following a maximum inspiration. Tidal volume (0.5 l) is the amount of air moved in and out of the lungs at each breath. The inspiratory reserve volume (3.3 l, 1.9 l) is that volume of air, taken in on a maximum inspiratory effort, over and above the tidal volume. The expiratory reserve volume (1.0 l, 0.7 l) is that volume of air expired by an active expiratory effort after the tidal volume has been expelled. That volume of air remaining in the lungs after a maximum voluntary expiration is the residual volume (1.2 l, 1.1 l). Finally, the functional residual capacity is that volume of gas remaining in the chest after normal passive expiration (2.2 l, 1.8 l) and is thus the sum of the residual volume and expiratory reserve volume. Most of these volumes may be measured using a spirometer. The residual volume, however, has to be estimated from determination of the dilution of a gas such as helium. On ageing there is a loss of the lung's elastic pull on the thorax, with a resultant increase in residual volume. FFFTF

MCQ 66

The physiological dead space of the lungs

1. represents the physiological shunt.
2. and anatomical dead space have dissimilar values in most circumstances.
3. is about 5 per cent of the functional residual capacity.
4. is greater in the erect than in the supine position.
5. decreases when circulating adrenaline concentrations increase.

Pulmonary ventilation is the amount of air inspired per minute. Functionally ventilation can be divided into alveolar ventilation, which is the volume of air reaching the alveoli where respiratory gas exchange occurs, and the dead space ventilation, that volume of air not exchanged with the blood. The anatomical dead space comprises the volume of the air passages proximal to the respiratory bronchioles. The total or physiological dead space comprises the anatomical dead space and the alveolar dead space, which is the volume in that portion of the lung containing poorly functioning alveoli. Neither the alveolar nor the physiological dead space has morphological correlates. On expiration gas from the alveoli, including carbon dioxide, becomes mixed with gas from the dead space, which contains no carbon dioxide. The volume of the dead space may be calculated from the following equation:

$$\text{vol. of air in alveoli} \times \text{alveolar } P\text{CO}_2 = \text{tidal vol.} \times \text{expired } P\text{CO}_2$$
$$(\text{tidal vol.} - \text{dead space}) \times \text{alveolar } P\text{CO}_2 = \text{tidal vol.} \times \text{expired } P\text{CO}_2$$

If the tidal volume = 500 ml, x is the dead space volume, $P\text{CO}_2$ (i.e. arterial $P\text{CO}_2$) = 40 mmHg and expired air $P\text{CO}_2$ = 25 mmHg, then

$$(500-x) \times 40 = 500 \times 25$$
$$x = 150$$

If the functional residual capacity is about 3000 ml then

$$\text{Dead space} = 150/3000 \times 100 = 5 \text{ per cent}$$

The dead space is greater in the upright position because of the effect of gravity. In an upright position the alveoli at the apices are relatively poorly perfused, so that the effective dead space is increased. Adrenaline dilates the lower airways by relaxing the smooth muscle and also dilates the upper airways by vasoconstriction, which reduces the volume of the nasal mucosa. The term physiological shunt describes the admixture in the pre-arterial blood of poorly oxygenated blood from the anatomical shunt pathways or from the pulmonary capillaries that are poorly ventilated. FFTTF

If a healthy person inhales dust particles they

1. may be partly cleared in the lungs if less than 10 μm in diameter.
2. can be cleared only by the ciliary action if they settle in the bronchi.
3. may be caught in the mucus if they reach the alveoli.
4. may be engulfed by macrophages in the alveoli.
5. will be cleared into the lymph vessels if they reach the alveoli.

The dead space of the respiratory tract serves to protect the rest of the tract by warming, moistening and cleansing the air. Inspiration is of sufficient duration for air to be very close to room temperature when it reaches the lungs. This warming is very efficient as air inspired at 6 °C is at 30 °C by the time it reaches the back of the nose. Expired air is 100 per cent saturated with water vapour. Most of the water vapour is added in the nose and the remainder is added in the lower airways. Saturation of the air with water vapour is important to remember in calculating volumes or pressures of samples of expired air. For example, the volume of expired air corrected to standard temperature and pressure (STP) is given by

$$\text{Volume}\,(V) \propto \frac{\text{Temperature}\,(T)}{\text{Pressure}\,(P)}$$

$$V_{\text{STP}} = V \times \frac{273}{(273+t)} \times \frac{P-\text{SVP}}{760}$$

where 273 is the absolute temperature, t the room temperature, P the barometric pressure and SVP the saturated water vapour pressure. The conducting airways are equipped to clear the air through the rhythmic movement of the cilia of the columnar epithelium. The cilia beat about 20 times per minute and carry particles along the trachea at a rate of about 0.4 mm s^{-1}. Mucus and foreign bodies are propelled to the main bronchi and trachea and are expelled by coughing. Small particles, which may reach the lower respiratory tract, are trapped by the mucus and removed by the phagocytic activity of the free macrophages in the alveoli. Larger objects which enter the airways trigger the cough reflex. Electrical, chemical or mechanical irritation of the nasal mucous membranes also causes closure of the larynx, constriction of the bronchi and slowing of the heart, a reflex related to the diving reflex. This helps to prevent the inhalation of liquids and noxious gases. TFFTT

MCQ 68

Increased airway resistance

1. is reflected in a reduced forced expiratory volume.
2. is seen on inspiration as compared with expiration.
3. is produced by the outward pull of the lung parenchyma.
4. is seen in response to histamine.
5. is seen in response to stimulation of irritant receptors.

The work of respiration includes not only the effort in overcoming the elasticity of the lungs and chest and the frictional resistance in the tissues, but also that involved in overcoming airway resistance. Similar factors affect airway resistance to those which affect resistance in the cardiovascular system. Airway resistance can be calculated by alveolar pressure/airflow. It is directly proportional to the size of the interactions between the flowing molecules and to the length of the airway, and inversely proportional to the fourth power of the radius. The airway diameter varies greatly with the condition of the intramural smooth muscle. Bronchiolar tone can be influenced by humoral agents such as histamine, which causes constriction, and by the sympathetic and parasympathetic nerve receptors. The pulmonary parenchyma pulls outwards on the lungs and this tends to dilate the airways. This distending action has been termed radial traction. Airways are also effectively distended by the transmural airway pressure, the intrapleural pressure normally being 5–8 cm water less than the airway pressure. The intrapleural pressure is less negative on expiration, especially during forced expiration when it becomes positive, so that airway resistance is greater in expiration than inspiration. Airway resistance is also increased in response to stimulation of irritant receptors. There are a number of ways of testing lung function. Lung volumes and capacities may be measured, and also the timed vital capacity. For the latter test the subject exhales as rapidly and fully as possible after a maximal inspiration. The volume of air expired in the first second is called the forced expiratory volume. An alternative is to use the peak-flow meter. Low values are found when there is an obstacle to air flow, as seen in asthma and other types of bronchoconstriction. TFFTT

MCQ 69

During normal quiet expiration

1. the external intercostal muscles contract.
2. the abdominal muscles contract.
3. the diaphragm contracts.
4. the larynx narrows.
5. the ribs move downwards.

On inspiration various muscles contract to enlarge the thoracic cavity and hence expand the lungs, as the visceral pleura cannot separate from the parietal pleura on the thoracic wall. Inspiration is an active process, about 75 per cent of the change in thoracic volume being produced by contraction and downward movement of the diaphragm, which is supplied by the phrenic nerve arising from cervical roots C_3 to C_5. The external intercostal muscles supplied by the thoracic intercostal nerves produce the rib movements. When they contract they elevate the lower ribs, which push the sternum outwards. Thus the chest wall moves upwards and outwards. When increased inspiratory effort is required, as in shortness of breath, the accessory muscles become important. The main ones are the scaleni, which elevate the first two ribs, and the sternomastoids, which raise the sternum and increase the anteroposterior diameter of the thorax. In deep breathing the extensors of the vertebral column may also contribute to inspiration. The abductor muscles in the larynx contract early in inspiration, pulling the vocal cords apart and opening the glottis. Expiration is a passive process brought about by the elastic recoil of the lungs and chest wall on relaxation of the inspiratory muscles. When a forced voluntary expiration is made such as blowing hard, expiratory muscles are used, as they are when the chest volume is taken below the resting level in a prolonged expiration. The expiratory muscles include those of the abdominal wall and the internal intercostals. FFFTT

MCQ 70

The intra-alveolar pressure

1. may be measured with an intra-oesophageal balloon.
2. is always subatmospheric.
3. depends on the airway resistance.
4. is usually less negative than the intrapleural pressure.
5. is lower than the intrabronchial pressure on expiration.

Intra-alveolar pressure and airway resistance are related. As with the flow of blood, the flow of air through the respiratory system is given by the equation

$$\text{flow} = \frac{\text{pressure gradient}}{\text{resistance}}$$

Similar factors also affect the airway resistance, namely the interaction of the flowing molecules, the length of the airways and the airway radius. However, the resistances are so small that flow can be produced by very small pressure gradients. Normally there is just a thin layer of fluid between the lungs and chest wall, so that the lungs will slide over the chest wall but will not pull apart. At the end of quiet expiration the tendency for the lungs to pull away from the chest wall is balanced by the pull of the chest wall. The action of these forces renders the pressure between the lungs and the chest wall subatmospheric and it remains so throughout a normal respiratory cycle, in contrast to the intra-alveolar pressure. As the inspiratory muscles contract and the size of the thoracic cage increases, the intrapulmonary pressure falls to −6 mmHg (−0.8 kPa). The resultant expansion of the lungs produces a slightly negative intra-alveolar pressure and air flows into the lungs, returning the pressure to atmospheric. On expiration, as the thorax returns to its resting volume the intrapleural pressure becomes less negative, falling to about −2.5 mmHg (−0.34 kPa) and the pressure within the alveoli rises to about 1 mmHg (0.13 kPa), so that air flows out of the lungs. Greater changes in pressure can be observed. Intrapleural pressure can fall to as low as −30 mmHg (−40 kPa) during a strong inspiratory effort and can increase to 52 mmHg (6.9 kPa) on a forced expiration. Intrapleural pressure can be determined by inserting a needle connected to a manometer between the two pleural layers. A close approximation to intrapleural pressure may be obtained under normal circumstances by measurement of the intra-oesophageal pressure. At constant volume and with the airways unobstructed the pressure in the alveoli can be recorded using a nasal catheter. FFTTF

MCQ 71

Surfactant in the lung

1. is a phospholipoprotein.
2. helps prevent over-inflation of the alveoli.
3. when deficient results in pulmonary distress syndrome in the newborn.
4. is increased when pulmonary blood flow is interrupted.
5. has the maturation of the synthetic mechanisms facilitated by cortisol.

By the law of Laplace, the pressure in a sphere $P = 2T/R$, where T is the tension in the walls and R is its radius. Because the alveoli vary in size by some three or four times, the pressure required to keep them inflated would also vary. There is obviously uniform pressure throughout the alveoli, a situation which is possible because of the surfactant which lines them. Surfactant also permits the lungs to be more easily inflated, increases compliance and has progressively less effect the more the lungs are inflated, which tends to prevent over-inflation of the alveoli. Surfactant is a phospholipoprotein which lowers the surface tension in the alveoli by forming a monolayer at the interface between the fluid lining the alveoli and the air. It is thought to be produced by large epithelial cells in the alveoli: the type II alveolar cells or pneumocytes. Insufficient surfactant in the newborn results in respiratory distress syndrome or hyaline membrane disease. Lack of surfactant causes alveoli to become unstable and collapse (atelectasis) at low lung volumes and the high surface tension results in oedema, leading to impaired gaseous exchange. The hypoxaemia, carbon dioxide retention and acidosis in turn lead to damage of the capillary endothelium in the lungs. Maturation of the synthetic mechanisms is facilitated by cortisol, the administration of which to pregnant women combats the disease in premature infants. Surfactant is reduced when pulmonary blood flow is interrupted, which may account for the atelectasis following a pulmonary embolus. Good expansion of the lung is necessary to spread the film over the alveolar surface, which is why after anaesthesia patients are encouraged to breath deeply. TTTFT

MCQ 72

Compliance of the lungs

1. is the change in volume produced by a unit change in distending pressure.
2. is increased if surfactant is depleted.
3. is influenced by scar tissue in the lungs (pulmonary fibrosis).
4. in normal respiration in the upright posture is greater at the apex of the lungs.
5. depends in part on airway resistance when measured during continuous breathing.

Compliance is a measure of the elasticity of the lungs and indicates the ease with which they can be distended. It is defined as the magnitude of the change in lung volume ΔV caused by a given change in the pressure difference across the lung wall ΔP. Compliance should be measured when there is no airflow, i.e. when the only determinant of pressure is the recoil of the lung. The pressure gradient is the difference between the pressure in the alveolar space and that in the pleural space. If respiration is stopped with the glottis open the former is equivalent to the barometric pressure. The latter is given by the pressure in the oesophagus, which is in essence a flaccid tube exposed to pleural pressure. If inflation is carried out in steps then a plot of transpulmonary pressure against volume is obtained, the slope of which gives lung compliance. The compliance of the human lung is about 0.2 l cmH_2O^{-1} or 2.0 l kPa^{-1}. It should be noted that because of the elastic recoil of the lung the compliance curve obtained during inspiration is different from that obtained during expiration. Lung volume at any pressure is greater during deflation than during inflation. This is known as hysteresis. The compliance *in vivo* will depend on airway resistance as this influences the pressure volume loop and the slope of the line joining the ends. The compliance is less at the apex. Because of the effect of gravity, the alveoli here are relatively more stretched at the end expiratory volume, so that they are less easily filled. Compliance depends not only on the elastic properties of the lung but also on the surface tension of the lungs and hence the surface of the lungs, as this will influence the tendency of the lungs of contract when stretched. Expansion of the lungs is restricted and hence compliance reduced in pulmonary congestion and with deposition of collagen in the walls of the alveoli (fibrosis). Compliance is increased in emphysema because of the loss of pulmonary elastic tissue. TFTFT

MCQ 73

Carbon dioxide is carried in the circulation

1. at a partial pressure of 40 mmHg in the arterial blood.
2. predominantly as carbonic acid.
3. in combination with haemoglobin.
4. in association with plasma proteins.
5. largely in the red cells.

Carbon dioxide is carried in the venous blood at a partial pressure of 46 mmHg (6.1 kPa) and in the arterial blood at 40 mmHg (5.3 kPa). The pressure gradient or difference in content between arterial and venous blood is thus very much less for CO_2 than for O_2. Carbon dioxide is carried in the blood from the tissue to the lungs in three main ways: 3.5 ml 100 ml^{-1} blood is carried in solution, 3.7 ml in combination with haemoglobin and 44.8 ml as bicarbonate. Carbon dioxide reacts with water to form carbonic acid which dissociates into bicarbonate and hydrogen ions. The formation of carbonic acid in plasma is quite slow and could not account for all the bicarbonate carried in the blood. It is chiefly formed in the red cells, where the reaction is catalysed by carbonic anhydrase. The hydrogen ions produced on dissociation of carbonic acid are buffered by the haemoglobin within the red cell. The capacity for this is increased as CO_2 enters the blood in the tissues, as reduced haemoglobin has a greater affinity for H^+ ions. The H^+ ions may also bind to plasma proteins. The bicarbonate ions diffuse out of the red cell, thus disturbing the distribution of electrical charges across the red cell membrane, so other anions—chiefly chloride— enter the cell, a process known as the chloride shift. This is aided by the presence of a chloride/bicarbonate carrier system in the erythrocyte membrane. Some of the carbon dioxide entering the red cell forms a loose complex with the terminal amino group of the reduced haemoglobin; some combination can occur with other proteins. TFTFF

MCQ 74

A shift of the oxygen–haemoglobin dissociation curve to the left is found with

1. decreasing temperature.
2. an increase of hydrogen ion concentration.
3. an increase in carbon dioxide tension.
4. increased 2,3-diphosphoglycerate.
5. anaemia.

A plot of the oxygen carried by the haemoglobin against the partial pressure of oxygen—the oxygen–haemoglobin dissociation curve—is sigmoid in shape. This is because the four oxygen molecules which bind to haemoglobin have differing affinities, the first oxygen molecule to bind having the lowest affinity and the last the highest. The shape of this curve is of some physiological significance. First, because the percentage saturation of haemoglobin varies little with changes in oxygen tension down to 60 mmHg, relatively high arterial oxygen content can be maintained even if there is a reduction in lung function. Second, the very steep change in the percentage saturation in the middle part of the curve (60–20 mmHg) allows oxygen release in the tissues in response to small changes in tension. A shift of the curve to the left means that a low partial pressure of oxygen is required for a given amount of oxygen to be bound, i.e. the haemoglobin has a greater affinity for oxygen. The oxygen dissociation curve is affected by carbon dioxide tensions and pH, by temperature and by the diphosphoglycerate concentration in the red cell. The curve is shifted to the left by a decrease in carbon dioxide (the Bohr effect), an increase in pH, decreased temperature or a fall in 2,3-diphosphoglycerate. The reverse changes shift the curve to the right, which is important since these changes occur in the tissues and hence allow oxygen release. Hypoxia causes an increase in 2,3-diphosphoglycerate and hence at altitude more oxygen is available in the tissues. In the fetus the oxygen dissociation curve is to the left of that of the mother, showing that fetal haemoglobin has a greater oxygen-carrying capacity. This is of importance if the fetus is to obtain oxygen from the mother. The myoglobin–oxygen dissociation curve is hyperbolic in form with no significant release of oxygen occurring before the partial pressure reaches 20 mmHg. Myoglobin thus acts as a temporary oxygen store. TFFTF

MCQ 75

Contributing to the functioning of the peripheral chemoreceptors is the fact that

1. they have a blood flow per unit mass similar to that of the brain.
2. they are influenced by changes in blood pressure.
3. they are more influenced by arterial PO_2 than by arterial oxygen content.
4. they increase in size and sensitivity if the body is continuously stressed by hypoxia.
5. they may be stimulated by a rise in blood hydrogen ion concentration.

Respiration allows for the control of the PO_2, PCO_2 and the pH of the fluids of the body. As part of the system controlling respiration are two types of respiratory chemoreceptors: the central chemoreceptors, sensitive to changes in CO_2 and H^+ ion concentrations; and the peripheral chemoreceptors in the carotid and aortic bodies, largely responsive to changes in PO_2. The central chemoreceptors are located on the ventral surface of the medulla, being situated in the interstitial fluid between the capillary supply and the cerebrospinal fluid. Effectively CO_2 acts on the respiratory centre by providing a rapid means of altering the hydrogen ion concentrations in the interstitial fluid bathing the bulbar neurones. The peripheral chemoreceptors in the carotid bodies lie near the bifurcation of the common carotid artery and in the aortic bodies. The epithelioid cells have rich innervation and the chemoreceptor afferents pass to the medulla in the IXth and Xth nerves. The blood supply per unit of tissue is very high, being greater than to any other tissue in the body. The peripheral chemoreceptors respond to reduced PO_2; they appear to be more influenced by arterial PO_2 than by arterial O_2 content, so are not influenced by a low haemoglobin level. At a normal PO_2 there is activity, the firing rate increasing noticeably below 60 mmHg, stimulating respiration. The chemoreceptor firing rate increases during hypotensive hypotension, not owing to a fall in blood pressure but rather to stagnant hypoxia. The peripheral chemoreceptors are also sensitive to changes in PCO_2 and pH, but provided there is no hypoxia the stimulation of respiration seen in response to elevated CO_2 and pH is not essentially due to peripheral chemoreceptors, but due to the central effect. FFTTT

MCQ 76

The medullary respiratory centre

1. comprises anatomically distinct inspiratory and expiratory centres.
2. activates the phrenic motor neurones.
3. ceases rhythmic activity if both vagus nerves are cut.
4. is unaffected by shifts in hydrogen ion concentrations in the body.
5. receives afferent inputs from pulmonary mechanoreceptors.

Respiration not only regulates carbon dioxide and oxygen tensions in the body, but also balances ventilation to the prevailing metabolic situation and modifies breathing rhythms to allow speaking and swallowing. Inflation of the lungs also needs to be kept within narrow limits. Respiration therefore requires a complex series of feedback mechanisms with many receptors and a centre for integration. The respiratory centre comprises interconnected neurones in the medulla and pons (pneumotaxic centre). At one time it was believed that there were separate inspiratory and expiratory centres. It is now thought that there are two regions in the medulla associated with respiration: the area of dorsal nucleus of the tractus solitarius, comprising inspiratory neurones; and the ventral nucleus retroambigualis and ambiguus, comprising inspiratory and expiratory neurones (50:50). Rhythmic activity originates in the nucleus of the tractus solitarius, containing α and β cells which feed back on each other. A given threshold level of activity is required in the α cells before they switch off. Negative inputs on these cells come from the vagus via the β cells and the pons. The centre controls the motor neurones of the respiratory muscles, primarily the diaphragm, supplied by the phrenic nerve. Inputs are received from the peripheral and central chemoreceptors and the centre can be directly influenced by oxygen, being depressed by very low tensions of oxygen. Other afferents may come from (as already indicated) pulmonary stretch receptors, baroreceptors, pain receptors and peripheral thermoreceptors. Monitoring the tidal volume depends on the load-detecting reflex of the thoracic muscle spindles. In addition to pulmonary stretch receptors, two other types of receptors exist with afferent pathways in the vagus: the irritant receptors and J receptors, which seem to respond to interstitial oedema. There are also hormones such as adrenaline and progesterone which cause an increase in respiration. Finally there is the voluntary control system. The voluntary system sends impulses to the neurones of respiratory muscles via the corticospinal tract, whereas the autonomic system supplies the same group of neurones via the reticulospinal tract. FTFFT

MCQ 77

Periodic or Cheyne–Stokes respiration may accompany

1. hyperventilation.
2. inspiration of pure O_2.
3. diabetes mellitus.
4. ascent to high altitude.
5. overventilation carried out with air containing 5 per cent CO_2.

Cheyne–Stokes breathing is a type of periodic breathing in which after a few breaths there is a break in respiration, or apnoea, followed by a period of deep breathing and the cycle is repeated. It can be produced experimentally by hyperventilation. Under normal conditions blood $P\text{CO}_2$ tension dominates the control of breathing, but after hyperventilation the blood $P\text{CO}_2$ tension falls below the minimum necessary for rhythmic breathing. If hyperventilation is carried out with air containing 5 per cent CO_2 no apnoea is seen. During the period of apnoea CO_2 accumulates in the body and an O_2 deficit develops, both of which stimulate respiration. If pure O_2 is inspired during hyperventilation the period of apnoea is prolonged. Gradually normal gas tensions are achieved and normal respiration is resumed. This pattern of respiration may be seen during sleep in some individuals at altitude and is caused by a low $P\text{O}_2$ and an altered pattern of central respiratory activity in sleep. A lowered $P\text{O}_2$ tension influences the ventilatory response to CO_2, so that at low partial pressure there is very little drive to ventilation. Then at relatively high partial pressures of CO_2 there is a sharp rise in the ventilatory response. This means that in contrast to voluntary hyperpnoea ascent to high altitude leads to maintained periodic breathing. Cheyne–Stokes respiration is also seen in disease states, most commonly in congestive heart failure and kidney failure or uraemia. The response may be due to increased sensitivity to CO_2 or, possibly in the case of cardiac disease only, to increased lung-to-brain circulation time. This means that it takes longer for changes in the arterial gas tensions to affect the respiratory centres. Another pattern of breathing, Kussmaul breathing, characterized by very deep breaths is produced to compensate for metabolic acidosis. It is this form of breathing which is seen in diabetes mellitus. TFTTF

MCQ 78

A low ventilation perfusion ratio

1. is seen in the apex of the lung as compared to the base when a person is in the upright posture.
2. indicates shunting.
3. will increase the systemic arterial partial pressure of oxygen (PO_2).
4. will increase the local carbon dioxide tension (PO_2).
5. will affect arterial PO_2 and PCO_2 to the same extent.

The ratio of alveolar ventilation to pulmonary blood flow is called the ventilation perfusion ratio. Given that total alveolar ventilation is approximately 3.75 l min^{-1} and the blood flow through the lungs 5 l min^{-1}, the overall ventilation perfusion ratio is 0.8. Because of the effects of gravity, when a person is in the upright position the blood vessels at the base of the lung are relatively dilated compared with those at the apex and the alveoli relatively less distended although they are ventilated more. Ventilation differences, however, are relatively minor as compared with the blood flow differences, so that at the apex alveolar ventilation is relatively great with respect to blood flow, giving a ventilation perfusion ratio of 3. At the base the reverse is true and the ratio is 0.5. A low ventilation perfusion ratio is also indicative of shunting, as seen with congenital cardiovascular abnormalities in which not all of the blood passes to the lungs with each circulation. A similar situation arises if blood flows through a large region of unventilated alveoli. The blood returning from this area will contain reduced haemoglobin, which in turn will result in a lowering of arterial oxygen content. This cannot be corrected by increasing ventilation as blood from well-perfused areas cannot carry any more oxygen. In contrast, since the carbon dioxide dissociation curve is more linear than that for oxygen, there can be some compensation from areas of high ventilation perfusion. In addition stimulation of respiration will help to reduce the arterial partial pressure of carbon dioxide. In contrast, if well-ventilated regions of the lung have a poor blood flow the carbon dioxide content will rise, as this situation will be equivalent to increasing the dead space. FTFTF

MCQ 79

Mouth-to-mouth respiration

1. is a form of positive pressure respiration.
2. has the disadvantage that expired air has a high carbon dioxide content.
3. has the limitation that the oxygen content is relatively inadequate.
4. is superior to mouth-to-nose ventilation.
5. should be performed at a rate of 15–20 inflations per minute.

The physiological mechanism of breathing is negative pressure respiration with air entering the lungs as a result of the negative pressure in the thorax. However, positive pressure respiration can also be used in artificial ventilation when the normal mechanisms are not functioning, as for example after administration of anaesthesia or after a head injury. It is also required after emergencies such as drowning, cardiac arrest, electric shock and carbon monoxide poisoning. It is important that artificial respiration be commenced as soon as possible as the oxygen content of the blood falls rapidly, being zero after 8 min, and damage to cortical cells of the brain occurs within 2–3 min. Mechanical respirators are of two types: cabinet respirators and positive pressure respirators. Cabinet ventilators, in which the pressure in the lungs remains atmospheric and negative pressure is applied to the outside of the thorax, are seldom used now. Most mechanical ventilators depend on intermittent positive pressure ventilation in which the lungs are ventilated by rhythmically blowing air into them via a tube inserted into the trachea, either directly or through the nose or mouth. A negative pressure phase may also be incorporated into expiration. If no mechanical pump is available, expired air resuscitation may be employed. The mouth and throat should first be checked and the head extended to ensure a clear airway. In mouth-to-mouth respiration the patient's nose is closed and air is blown into the patient's mouth until the chest is seen to rise. Passive exhalation is then allowed. The process is repeated 10–12 times per minute. Mouth-to-nose ventilation, which is considered superior as full extension of the neck allows the airway to be opened, can be combined with full closure of the mouth. In addition inflation of the stomach is less likely. Even though the expired air of the operator contains only 16 per cent oxygen, it is possible to maintain an adequate oxygen supply to the patient. The high carbon dioxide content stimulates the respiratory centre and accelerates initiation of spontaneous respiration. TFFFF

MCQ 80

Breathing oxygen at high pressure may cause

1. a reduction in cardiac output.
2. apnoea.
3. brain damage.
4. retinal damage.
5. carbon monoxide to be driven off more quickly than normal.

Many conditions of hypoxia can be ameliorated by oxygen therapy. The patient may be given gas containing a high proportion of oxygen or hyperbaric oxygen, for which the patient is treated in a pressure chamber with oxygen at elevated pressure. Mixtures containing 60 per cent oxygen may be breathed indefinitely, but pure oxygen or oxygen above atmospheric pressure can only be breathed by man for periods of hours, otherwise damage results. Oxygen at high tensions produces a small increase in oxygen saturation, leading to a small fall in pulmonary ventilation but not apnoea. Hyperbaric oxygen drives off carbon monoxide from blood more rapidly than normal. There is a direct effect on the lungs, with respiratory tract irritation, nasal congestion, sore throat, coughing and decreased surfactant production. Because of the increased vagal tone, cardiac output falls and blood flow through the brain and kidneys is reduced. There is constriction of the cerebral vessels and in the longer term there is brain damage, with dizziness, convulsions and finally death. Newborn infants treated with oxygen for prolonged periods have been found to have retinal damage, which in some cases has led to blindness. At the biochemical level high oxygen pressure affects the sulphydryl groupings of enzymes in the Krebs cycle so that the supply of high-energy phosphate fails and the metabolic functions of the cells suffer. TFTTT

MCQ 81

The renal glomeruli in man

1. are found only in the cortex.
2. always have an afferent arteriole with a narrower lumen than the efferent arteriole.
3. are part of a portal system of vessels.
4. have fenestrated capillaries.
5. have blood filtrate barriers about 10.0 nm thick.

The kidneys are two bean-shaped organs each weighing 150 g in man. The indentation on the medial border is the hilum, from which passes the ureter and through which pass the renal artery and vein. If the kidney is cut longitudinally it is seen to comprise an outer cortex and an inner medulla made up of pyramids. The functional unit of the kidney is the nephron, of which there are about 1.2×10^6. The nephron arises as a blind-ended tube and can be divided into three main regions: the proximal convoluted tubule, the loop of Henle and finally the distal convoluted tubule, which empties into the collecting duct. The dilation at the blind end of the tube is filled with a tuft of capillary loops, the glomerulus. This is formed by branching of the afferent arteriole into interconnecting capillary loops. Blood leaves by the efferent arteriole which in the case of cortical nephrons breaks up into a network surrounding the tubule. The glomeruli are thus part of a portal system of vessels. The efferent arterioles have a luminal diameter similar to that of the afferent vessels, but possess a thinner wall. Every day about 180 l water and contained solutes are filtered in man through the glomeruli. The structure of the capillary wall and the adhering capsule determines the composition of the filtrate. The glomerular membrane through which the plasma is filtered comprises three layers: the fenestrated capillary endothelium, the basement membrane, 300 nm thick, and finally the epithelium, which possesses podocytes. The latter are said to form split pores which allow free passage of water and small molecules, but restrict the passage of colloid. TFTTF

MCQ 82

The human juxtamedullary nephrons

1. comprise over 30 per cent of the nephrons in the kidney.
2. each have a short loop of Henle.
3. are restricted to the renal medulla.
4. have afferent arterioles which break into a profuse peritubular capillary network on leaving the glomerulus.
5. generally have a higher glomerular filtration rate than the cortical nephrons.

The juxtamedullary nephrons comprise about 10–20 per cent of the total, the cortical nephrons comprising the rest. The glomeruli of both types of nephron lie in the cortex, but those of the juxtamedullary nephrons, which are larger, are situated in the deeper layers of the cortex. The proximal tubules of these latter nephrons are longer and the loops of Henle, which are very long with a thin as well as a thick ascending limb, penetrate the inner medulla. In contrast the cortical nephrons have a short loop of Henle which extends into the outer zone of the medulla. The afferent arterioles of the juxtamedullary glomeruli are short and straight and rapidly break up into the vasa recta, which are specialized peritubular capillaries. There is, as described in MCQ 87, an osmotic gradient across the kidney which could be dissipated by blood flowing through the kidney. However, the hairpin arrangement of the vasa recta means that they act as countercurrent exchangers. Thus sodium, etc., are taken up by simple diffusion at the arterial end of the capillary, so that the osmotic pressure increases as the blood flows towards the loop and falls again as the capillary leaves the region of high osmolality. The glomeruli of the juxtamedullary nephrons have a higher glomerular filtration rate, which is reduced during a water diuresis and increased with sodium intake, and the nephrons may play an important part in sodium balance. There is a significant positive correlation between the number of juxtamedullary loops and the ability of the animal to concentrate its urine. FFFFT

MCQ 83

Under normal conditions at the glomerulus

1. the pressure in the capillaries is higher than in most other capillaries.
2. the pressure in the Bowman's capsule is approximately 18 mmHg.
3. the net filtration pressure is about 10 mmHg.
4. the capillaries are more permeable to water than others in the body.
5. the afferent arterioles offer more resistance to blood flow than the efferent.

The structure of the glomerular membrane described in MCQ 81 means that a much larger fraction of water and small molecules leaves the glomerular capillaries than other capillaries in the body; indeed, the capillary permeability is some 100 times greater. Molecular sieving occurs in the membrane; whether or not a molecule is filtered depends on its molecular weight and also, to a degree, on its shape and charge. Haemoglobin, of molecular weight 67000, is at the limit of filtration and the mean pore radius of the filter is around 3.4–4 nm. The forces driving filtration are similar to those governing exchange across capillaries in general. The effective filtration pressure is the gradient of hydrostatic pressure across the glomerular membrane minus the colloid osmotic pressure of the capillary blood. The effective filtration is greater than that across normal capillaries. The high constant pressure is maintained as the efferent arterioles offer more resistance than the afferent. The adrenergic constrictor activity of the afferent arterioles is important, although the adrenergic fibres to the efferent arterioles may play a role. If the afferent arteriole vascular tone were increased more than the efferent, then filtration would decrease. The converse is also true. Because the plasma proteins are not filtered, the plasma oncotic pressure increases along the glomerular capillary. As this happens the ultrafiltration decreases until, if a point is reached when the oncotic pressure balances the hydrostatic pressure, it ceases. TFTTF

MCQ 84

If a substance is filtered and actively reabsorbed from the renal tubule then

1. it may be used to determine the renal plasma flow.
2. its renal clearance is greater than that of inulin.
3. the ratio of its rate of urinary excretion divided by its plasma concentration may be the same as that for glucose.
4. its concentration in the distal tubule is higher than that in plasma.
5. its clearance value is equal to that of creatinine.

Once a substance has been filtered at the glomerulus it may pass out directly into the urine, it may be reabsorbed or further amounts may be secreted into the urine. When considering the amount of a substance removed from the body in a given time it is usual not to specify the amount appearing in the urine, but rather the volume of plasma which was cleared of that substance. Clearance is in essence a mathematical concept and is defined as the volume of plasma which is completely cleared of a substance in one minute. An equation for the calculation of clearance (C) may be obtained from the basic statement that the amount of a substance (S) appearing in the urine is equal to the amount leaving the blood. If U is the urinary concentration of S and V the volume of the urine, then the amount of S appearing in the urine is UV. If the concentration of S in the plasma is P then the amount cleared from the plasma is CP. Since $CP = UV$, then $C = UP/V$. If a substance is freely filtered but is not secreted or reabsorbed and neither metabolized nor stored in the kidney, then its clearance represents the volume of plasma filtered, i.e. the glomerular filtration rate (GFR). If such a substance is easily measured, then it may be used to measure GFR. Such substances include the exogenous polysaccharide inulin or the endogenous creatinine. If, in addition to being filtered, a substance is actively secreted, so that all the blood passing to the kidney is cleared of that substance, then its clearance rate is greater than that of inulin and equal to the renal plasma flow. If a substance is actively reabsorbed, as for example glucose, then it does not appear in the urine and hence its clearance is 0. FFTFF

MCQ 85

Renal blood flow

1. may be estimated from urea clearance.
2. is 20–25 per cent of cardiac output.
3. largely passes to the renal medulla.
4. is decreased during exercise.
5. always remains constant in the healthy individual.

As indicated in MCQ 84, renal plasma flow may be calculated by determining the clearance of a substance which is filtered and secreted in the kidney, as for example *para*-aminohippuric acid (PAH). The renal blood flow may then be calculated from the packed cell volume. It is very high compared to that of other organs, being about 200 ml min^{-1}, which represents about 20 per cent of the cardiac output. As a result of the high blood flow, the oxygen extraction of the kidney is quite low even though the oxygen consumption is high. Also, oxygen consumption varies with blood flow. Blood flow within the kidney is not homogeneous: the cortex is better perfused than the medulla, which receives only 7–10 per cent of the renal flow. The inner medulla receives only 1 per cent. A relatively slow medullary flow is of importance for the development of the hyperosmolality of the inner medulla, and hence for the production of hypertonic urine. In contrast to cortical flow, medullary blood flow does not show autoregulation. Autoregulation means that blood flow remains remarkably constant between blood pressures of 90 and 200 mmHg as a result of alterations in the calibre of the preglomerular resistance section of the afferent arterioles. Blood flow does not, however, remain constant under all conditions. During mild exercise, for example, it can fall by 30 per cent and profound renal vasoconstriction is seen during shock. This probably results from circulating adrenaline and noradrenaline. FTFTF

MCQ 86

The fluid in the proximal convoluted tubule

1. is generally hypertonic to plasma.
2. is rich in potassium.
3. has less bicarbonate than plasma.
4. has a flow rate which falls along the tubule.
5. has a similar pH to that in the ureter when the kidneys excrete an acid urine.

The proximal tubule is lined by high cuboidal epithelium and possesses invaginations, which together with the microvilli of the brush border create the extensive surface which increases the marked absorptive capacity of this part of the renal tubule. Of the 170 l filtered per day, 80 per cent is reabsorbed in the proximal tubule, whatever the degree of hydration, and hence this is called obligatory absorption. Since such a large volume of fluid is reabsorbed, the flow rate falls along the tubule. Sodium is actively reabsorbed from the proximal tubule and chloride follows to preserve electrical neutrality. Their concentration builds up in the lateral intracellular space and so water is drawn after. This can be demonstrated as the rate of water transport is always proportional to that of the solute, so that the fluid remains isotonic along the tubule. Other substances such as glucose and amino acids are actively reabsorbed and this too facilitates water reabsorption and thus the passive absorption of substances, such as urea, to which the tubular epithelium is permeable. Little is known of the mechanism of potassium reabsorption, but it could include both active and passive mechanisms. There are also active mechanisms for the uptake of calcium and phosphate and also bicarbonate and magnesium. Hydrogen ions enter the proximal tubule in exchange for sodium and combine with bicarbonate to form carbon dioxide, which diffuses back into the plasma. Although this represents over 90 per cent of excreted hydrogen ions, the system is more properly a mechanism for recovering bicarbonate ions, as the permeability of the cell membrane is too low to allow reabsorption of bicarbonate by simple diffusion. Bicarbonate behaves as if it had a threshold for excretion. FFTTF

MCQ 87

The properties of the loop of Henle which contribute to the formation of the concentration gradient across the kidney include

1. high permeability of the descending limb to electrolytes.
2. permeability of the descending limb to water.
3. active pumping of sodium chloride in the ascending limb.
4. active reabsorption of sodium in the descending limb.
5. high sensitivity to vasopressin.

There is a large osmotic gradient between the cortex and the medulla, which allows excretion of urine over a wide concentration range. The gradient is produced by the loop of Henle acting in association with the vasa recta as a countercurrent multiplier system. A number of properties of the loop of Henle contribute to the production of a highly concentrated interstitial fluid in the inner medulla, some 1400 mOsm kg^{-1}. Sodium and chloride are actively co-transported from the thick ascending limb of the loop of Henle. This process, which only occurs at this point, can achieve an osmotic gradient of some 200 mOsm kg^{-1}. The ascending limb is impermeable to water and the descending limb permeable to water, somewhat permeable to urea and nearly impermeable to ions. The thin portion of the ascending limb is relatively impermeable to urea and sodium chloride, but the thick portion is relatively permeable to these solutes. Thus sodium chloride leaves the ascending limb, so that the fluid becomes hypotonic and the surrounding interstitium becomes more concentrated. Because the descending limb is permeable the fluid in this portion also becomes more concentrated. This happens along the loop of Henle, so that an osmotic equilibrium is established at each level of the loop, with the tubular fluid becoming more concentrated as it descends to the medulla and more dilute as it leaves. The sodium continually enters the descending limb by diffusion, is carried round the limb and transported out again. Because of the isosmotic reabsorption of sodium and water in the proximal tubule the concentration of urea in the tubular fluid increases. It is subsequently reabsorbed from the collecting duct and enters the loop of Henle. It is thus recycled, which contributes to the large quantities of urea in the medullary interstitium. As indicated in MCQ 85 the vasa recta play a role in maintaining the gradient, but it is not believed to be subject to hormonal control. Vasopressin acts on the final part of the distal tubule and on the collecting duct. FTTFF

MCQ 88

The cells of the distal convoluted tubule

1. are simple squamous epithelial cells.
2. contain carbonic anhydrase.
3. adjust the composition of urine to its final form.
4. actively reabsorb potassium.
5. secrete hydrogen ions.

The distal convoluted tubule possesses cuboidal cells which have fewer microvilli than those of the proximal tubule and longer tight junctions between adjacent cells. While the majority of the filtrate is reabsorbed in the proximal tubule, some reabsorption occurs in the distal tubule, but this is regulatory, contributing to the final adjustment of the urine composition. Tubular fluid does not achieve its final composition in the distal tubule, as it has been found experimentally using micropuncture (a technique extensively used in renal investigations in experimental animals) that the fluid is isotonic with plasma. About 10 per cent of the filtered sodium (or about 2500 mmol) is reabsorbed in the distal tubule. Of this some 500 mmol is controlled by aldosterone. The tubular absorption of sodium is a process which results in the secretion of potassium or hydrogen ions. The tubular sodium load determines the extent of the negativity of the luminal membrane and hence the gradient for potassium secretion. Potassium ions are secreted passively via an electrochemical gradient. Thus any procedure which increases the existing electronegativity causes increased potassium secretion. Calcium excretion is also regulated in the distal nephron, the main factor involved being the plasma parathyroid hormone concentrations. Cells of the distal tubule also contain carbonic anhydrase, which acts as a catalyst for the conversion of carbon dioxide and water to carbonic acid, the latter dissociating into hydrogen and bicarbonate ions. This part of the renal tubule thus plays an important part in the excretion of hydrogen ions. The bicarbonate is reabsorbed and the hydrogen ions exchanged for sodium ions are excreted. The bicarbonate ions reabsorbed are not therefore the same ions present in the blood but are derived from carbon dioxide entering the renal tubule. FTFFT

MCQ 89

If a substance has T_{max} (transport maximum) this implies that

1. the main factor contributing to the quantity reabsorbed is the length of time the substance is in the tubule.
2. reabsorption is an active process.
3. a constant fraction of the substance will be reabsorbed.
4. reabsorption may be linearly related to concentration in plasma.
5. below the threshold level none of the substance should appear in the urine.

As indicated in MCQ 86 the majority of filtered water and solutes are reabsorbed in the proximal tubule. Reabsorption of ions and foodstuffs such as glucose is obviously of vital importance and specific mechanisms exist to facilitate the process. Some of these mechanisms have a limited reabsorptive capacity, in which case a transport maximum is said to exist. Substances exhibiting a transport maximum, such as glucose, amino acids, phosphate and sulphate, are all actively reabsorbed in the proximal tubules and cannot diffuse passively across the epithelium. The tight junctions provide fairly effective barriers to these materials. Substances such as sodium and potassium have no T_{max} since raising the concentrations sufficiently to saturate the transport mechanisms results in greater passive diffusion. Normally there is enough time during the passage of the filtrate along the tubule for all the glucose to be absorbed and thus tubular length is not a critical feature. This would, however, be important for gradient–time limited active reabsorption. As the plasma concentration of glucose is increased, so the amount filtered increases until a point called the renal threshold is reached, when some glucose appears in the urine. Thus the amount of glucose reabsorbed rises linearly as the plasma glucose increases up to the renal threshold (180 mg ml^{-1}, 10.0 mM). Thereafter the amount reabsorbed remains constant and the amount excreted increases with the plasma concentration. The reabsorption curve does not abruptly flatten off but has a splay, which results from gradual saturation of the transport mechanisms and the fact that all nephrons do not behave identically. Glucose reabsorption occurs, via a carrier, with sodium, the energy coming from the sodium gradient. The limitation to reabsorption is thus the number of available sites on the carriers.
FTFTT

MCQ 90

Urine with a pH of less than 5.0 has

1. an acidity that may lie within the normal range.
2. large quantities of free hydrogen ions.
3. hydrogen ions secreted by ion exchange with potassium ions (which are reabsorbed to maintain electrical neutrality).
4. a higher than usual concentration of bicarbonate.
5. a higher than average concentration of ammonium ions.

Urine may be excreted with a pH in the range 4.5–8.5, depending to a large extent on the nature of the diet. Metabolizing protein reduces the pH because of the phosphorus and sulphur contained. Any acid released can be buffered by bicarbonate and protein buffers. However, at some stage it must be excreted. Hydrogen ions enter the tubular fluid in both the proximal and distal parts of the tubule. Hydrogen ions may enter the tubule as part of the process of reabsorption of bicarbonate ions, as discussed in MCQ 86. In this case the hydrogen ions are not excreted in the urine but are reabsorbed or excreted as water. If, however, new bicarbonate is added to the plasma in maintenance of plasma pH, then hydrogen ions combine with the filtered hydrogen phosphate to form dihydrogen phosphate and hence the ions are excreted. In plasma the ratio of alkaline to acid phosphate is 4:1 and hence pH is given by

$$\text{pH} = 6.8 + \log_{10} \frac{[Na_2HPO_4]}{[NaHPO_2]} \quad \text{is } 7.4$$

In acid urine with a pH of 4.5 the ratio is 200:1. Hydrogen ions are excreted mainly in exchange for sodium, i.e. hydrogen ions compete with potassium ions in the exchange. Hydrogen ion excretion is thus increased by hypokalaemia and decreased by hyperkalaemia. Hydrogen ions may also be excreted in the form of ammonium ions. If a large hydrogen load is to be excreted then ammonium can be formed in the renal tubular cells largely from glutamate, the reaction being catalysed by glutaminase. Relatively smaller amounts of ammonia are formed from circulating amino acids by the action of amino acid oxidase. TFFFT

MCQ 91

The juxtaglomerular apparatus

1. comprises in part densely staining epithelial cells of the proximal convoluted tubule.
2. comprises in part granular filled smooth muscle cells of the efferent and afferent arterioles.
3. releases angiotensin II.
4. may contribute to the maintenance of systemic blood pressure.
5. secretion may be controlled by sympathetic nerves.

Before the distal tubule becomes convoluted it passes close to the glomerulus, coming in contact with the afferent and efferent arterioles. Three specialized types of cells are found here. First are the granular cells: the cells surrounding the afferent arterioles at the contact areas are filled with granules containing renin. Second are the cells of the extraglomerular mesangium, which make gap junctions with themselves and are located in the hilus between the afferent and efferent arterioles. Finally there are the macula densa cells, which arise from the epithelial cells of the distal tubule at the contact regions with the arterioles. These cells have some unusual biochemical properties so that there is, for example, a complete absence of Na^+/K^+-ATPase. The renin released from this region is a protease which acts on angiotensinogen, an α_2-globulin in the plasma, to release the decapeptide, angiotensin I; the latter is converted to angiotensin II in the lung by a converting enzyme. Angiotensin II is a powerful vasoconstrictor and is also involved in the release of aldosterone. Thus the juxtaglomerular apparatus plays a role in the maintenance of fluid balance, blood volume and blood pressure. Changes in blood pressure influence renin release, probably via a pressure- or stretch-sensitive region in the afferent arteriole. Renin release can be stimulated by a fall in plasma volume. It is suggested that this is via a chemosensitive mechanism which responds to a reduced delivery of sodium chloride to the distal tubule (as a result of reduced glomerular filtration rate). Renin release can also be stimulated neurally, probably via β-receptors. It is also subject to humoral influences. FTFTT

MCQ 92

After taking a water load of one litre

1. water is rapidly absorbed through the stomach.
2. thirst will not be relieved while water stays in the gastrointestinal tract unabsorbed.
3. there is reduced chemoreceptor activity.
4. there is reduced secretion of the posterior pituitary hormone vasopressin.
5. excess fluid will be excreted by the kidney over the following 24 h.

When a water load is ingested it is absorbed to keep the contents of the gastrointestinal tract isotonic with plasma (see MCQ 114). The main site of absorption of water and salt is the upper small intestine, although very small amounts may be absorbed through the stomach. Absorption of the water dilutes the extracellular fluid and ultimately the intracellular fluid. The reduced osmolality, acting via the osmoreceptors, relieves the sense of thirst. Thirst, however, may be transiently relieved by stimuli from the oropharynx, i.e. by reducing the dryness of the mouth and throat and by fullness of the stomach. The thirst centre is also influenced by the degree of cardiovascular filling and through cerebrocortical activity. The reduced extracellular fluid osmolality, again acting via hypothalamic osmoreceptors, reduces the secretion of vasopressin, but it is not clear if these are the osmoreceptors concerned with water intake. Certainly the two set points appear different, the osmolality at which thirst ceases being higher than the osmolality at which vasopressin secretion is no longer observed. This is important in preventing excessive increases or decreases in extracellular fluid osmolality. As vasopressin concentrations are reduced, so the permeability of the collecting duct to water decreases and the excess water is excreted. The majority of the water is excreted within 1–2 h. If, however, the individual is dehydrated when the water load is ingested then the water will be retained and the plasma osmolality returned to normal. FFFTF

MCQ 93

An osmotic diuresis may result from

1. lack of insulin.
2. ingestion of alcohol.
3. furosemide.
4. mannitol.
5. aldosterone.

A diuresis is an increased urine flow; it can take the form of a water diuresis or an osmotic diuresis or can be produced from inhibition of tubular transport processes. A water diuresis results from low or absence of circulating plasma vasopressin, so that reabsorption in the proximal tubule is normal, but water reabsorption in the distal tubule is reduced. The maximum urine flow that can be achieved is 16 ml min^{-1}. Alcohol is believed to suppress vasopressin secretion, thus producing a water diuresis. An osmotic diuresis is produced by a substance which is freely filtered, which shows little or no reabsorption in the renal tubule and has no pharmacological action on tubular transport mechanisms. Solutes not reabsorbed in the proximal tubule exert a very marked effect in preventing reabsorption of water because as the volume of tubular fluid falls their concentration rises. Reabsorption of water (and indirectly sodium) is inhibited and very large volumes of urine are produced. In patients with diabetes mellitus the plasma levels of glucose are so high that the filtered load exceeds the glucose T_{max} (see MCQ 89) and in addition ketone bodies such as acetoacetate and β-hydroxybutyrate may be present in excess of their T_{max}. Mannitol also possesses all the properties of an osmotic diuretic. The amount of solute in the urine can also be increased by inhibition of electrolyte transport by specific drugs. Such drugs fall into two main categories. The first comprises drugs such as furosemide and ethacrynic acid which directly block sodium chloride transport out of the thick ascending loop of Henle. Thus higher than normal concentrations of chloride and sodium are excreted, preventing reabsorption of water. Failure of the chloride pump also results in dissipation of the osmotic gradient across the kidney, so that highly concentrated urine cannot be excreted. The second group of drugs are the carbonic anhydrase inhibitors. As discussed in MCQ 88 this enzyme is necessary for bicarbonate reabsorption and when it is absent additional bicarbonate and hence water will be excreted. TFFTF

MCQ 94

Excretion of drugs by the kidney

1. may be an active process.
2. occurs largely in the proximal tubules.
3. includes clearance rates of 0–600 ml min^{-1}.
4. may be enhanced if the drugs are lipid-soluble.
5. will occur more rapidly if the drugs are non-ionic.

The excretion of drugs by the kidney depends on their chemical nature, but excretion is not usually greater if they are not ionized or are lipid-soluble. Excretion of drugs may cover the whole range of renal clearances. The proximal tubule has a relatively non-specific transport mechanism which secretes a variety of organic acids into the tubule. Such substances include ethereal sulphates, steroids and other glucuronides produced mainly in the liver as part of its detoxifying function. *Para*-aminohippuric acid (PAH), used to determine renal plasma flow, can compete with the 18-glucuronide and 3-glucuronide derivatives of aldosterone for secretion and therefore shares the same mechanism. The transport is enhanced by acetate and lactate and inhibited by probenecid and dinitrophenol—mercurial diuretics. Probenecid has been used clinically to inhibit excretion of penicillin, which is also transported by this system. Clinical advantage is also taken of this system in testing tubular function, as iodated radiopaque contrast agents such as diodrast are similarly excreted. A second mechanism exists in the proximal tubule, which secretes stronger organic bases such as guanethidine, thiamine, choline and histamine. As with the first system, the physiological significance of this mechanism is not clear. A third system has also been described which secretes ethylenediaminetetraacetic acid (EDTA). TTTFF

MCQ 95

In chronic renal failure

1. there is hypokalaemia.
2. the partial pressure of carbon dioxide in plasma tends to be low.
3. there is hypercalcaemia.
4. normocytic normochromic anaemia occurs.
5. protein in the diet should be restricted.

Renal insufficiency appears when the number of functioning nephrons falls below 30 per cent of normal or when the glomerular filtration rate is less than about 50 ml min^{-1}. At this stage there is noticeable retention of non-protein nitrogen such as urea, uric acid, creatine and creatinine. Acute renal failure, which commonly follows severe shock such as circulatory shock, occurs when the kidney suddenly becomes unable to maintain homeostasis. In those cases which recover, four stages are recognized: first anuria or oliguria, then the early diuretic phase, late diuretic phase and finally progressive recovery. In contrast, chronic renal failure develops progressively over a number of months or years and results from diseases of both kidneys as in, for example, chronic infection. Despite the damage to the kidney, the amounts of urea, creatinine, potassium and sodium excreted daily is little altered as a result of varying mechanisms. For example, constant excretion of creatinine in the face of a falling glomerular filtration rate is achieved by a rise in the plasma concentrations. Plasma levels of sodium and potassium remain constant since fractional excretion of electrolytes and water increases as the glomerular filtration rate falls. Ionized calcium in the plasma tends to fall because of the phosphate retention, the product of calcium and phosphate concentrations remaining constant. The plasma carbon dioxide tension is also low as ventilation is stimulated because of the metabolic acidosis which exists. Another manifestation of the condition is normocytic, normochromic anaemia, which results from inadequate bone marrow activity, probably as a result of erythropoietin deficiency. FTFTT

MCQ 96

In the urinary bladder

1. there is smooth muscle, the detrusor muscle.
2. the internal pressure increases linearly with volume.
3. a volume of approximately 500 ml produces the first urge to void.
4. there are tension receptors in the walls.
5. contraction of the smooth muscle can be brought about through stimulation of the parasympathetic fibres travelling in the pelvic nerve.

The walls of the bladder comprise smooth muscle—the detrusor muscle—arranged in spiral, longitudinal and circular bundles. The bladder epithelium consists of a superficial layer of flat cells and a deep layer of cuboidal cells. The ureters enter the bladder obliquely (to prevent reflux of urine) at the upper corners of the trigone, a triangular area of fine smooth muscle fibres. At the lower tip of the trigone the bladder leads into the urethra. Muscle bundles termed the internal urethral sphincter pass either side of the urethra, but can only relax during micturition with the assistance of the detrusor muscle. Further along the urethra is a sphincter of skeletal muscle, the external urethral sphincter, which can hold the urethra closed even against strong bladder contractions. If the pressure within the bladder is measured it is found that there is an initial increase on commencement of filling of the bladder, but then as filling continues the pressure remains constant until bladder capacity is reached. This depends in part on the properties of smooth muscle and in part on Laplace's law, which states that

$$\text{wall tension} = \text{internal pressure} \times \text{radius}$$

When the capacity of the bladder, some 250–450 ml, is reached the pressure rises steeply. The first urge to void is felt when the bladder contains 150–250 ml. There are stretch receptors in the bladder which activate parasympathetic nerves arising in the second to fourth sacral segments, so that bladder contraction results. The somatic nerves to the external sphincter are simultaneously inhibited, resulting in urination. This process can be delayed at will through activity of the descending pathways from the cerebral cortex which inhibit the bladder parasympathetic response. This control has to be learned during early childhood. TFFTT

Dietary carbohydrates

1. contain nitrogen.
2. may be stored in the body in the form of glycogen.
3. are solely polymers of hexoses.
4. are not essential constituents of the diet.
5. contribute about 6 per cent of the total calories of the British diet.

The six required components for an adequate diet are carbohydrates, fats, proteins, minerals, vitamins and water. A necessary component of food is also dietary fibre, comprising a heterogeneous group of polymeric carbohydrate compounds including celluloses, hemicelluloses and pectins and a non-carbohydrate unit, lignin. Food is required for two major purposes, the chief one being the provision of energy and the other the provision of structural material. Protein is an indispensable constituent of the diet as it is the only source of amino acids, including the essential amino acids which cannot be synthesized in the body. Dietary carbohydrate comprises largely, but not solely, polymers of hexoses and contains no nitrogen. In the normal diet carbohydrates provide more than 50 per cent of the energy content. If the amount of carbohydrate ingested is greatly reduced, ketosis may develop. As both carbohydrate and fat are largely valuable as sources of energy they can replace each other to a considerable extent, so long as a minimum of 10–14 g of fat is ingested. Animal fats are important sources of the fat-soluble vitamins A and D and it is also said that there are some essential fatty acids. Other fat-soluble vitamins are E, derived from cereals, and K from vegetables. The water-soluble vitamins are the B group, derived from yeast and meat, and the C group, present in fresh fruits and citrus fruits. The most important inorganic substances in the diet are calcium, sodium, potassium, magnesium, phosphorus, iodine and chloride. Dietary deficiency may occur especially in calcium, the daily requirement of which is 0.8 g, and iron, the requirement of which is 12.0 mg.
FTFTF

MCQ 98

Basal metabolic rate may be most markedly altered by

1. the exercise performed in reaching the laboratory, including climbing up several flights of stairs.
2. taking a snack 2 h before.
3. the fact that pure oxygen is inspired during the measurement.
4. apprehension.
5. a room temperature of 80 °C.

When food is oxidized in the body energy is released. For each class of foodstuff there is a characteristic respiratory quotient (RQ) or ratio of carbon dioxide produced to oxygen consumed and a given energy yield. The standard unit of heat energy was formerly the calorie, the amount of energy necessary to raise the temperature of 1 kg water by 1 °C, but more recently the kilojoule (kJ) has been adopted. The values may be determined using a bomb calorimeter. For pure carbohydrate the RQ = 1 and 1 l oxygen yields 20.9 kJ (5 kcal); for fat the RQ = 0.7 and the energy equivalent is 20.1 kJ (4.8 kcal). Indirect estimates give an RQ of 0.8 for protein. The metabolic rate is the amount of energy liberated per unit time, the basal metabolic rate (BMR) being that under standard resting conditions, namely at complete rest 12–14 h after a meal and in an equable environmental temperature. Under these conditions a certain amount of energy is used to maintain the activities of vital organs like the heart, kidneys, lungs, etc., but the greater part is converted to heat to maintain body temperature. Basal metabolic rate may be estimated by measuring the oxygen uptake using a spirometer. A value for RQ is generally assumed, a value of 0.8 generally being taken, and the number of calories produced for the oxygen used may be determined. Basal metabolic rate is more closely related to surface area than to height or weight, so that it is generally given in kJ m^{-2} h^{-1}. The surface area may be calculated from the Dubois formula, $A = W^{0.425} \times H^{0.725} \times 71.84$, where W is the weight and H is the height. The basal metabolic rate for a male aged 20 years is 41.4 kcal m^{-2} h^{-1} and for a female is 36.1 kcal m^{-2} h^{-1}. In addition to being influenced by sex, basal metabolic rate is affected by age, race, emotional state, climate, body temperature and circulating levels of thyroid hormones and catecholamines. TTFFF

MCQ 99

The specific dynamic action of

1. food leads to an increase in oxygen consumption after a meal.
2. food persists for a number of hours after a meal.
3. protein is 5 per cent of its total calorific value.
4. protein is partly due to the oxidative deamination of amino acids in the liver.
5. protein is less than that of fat.

Organisms break down the three classes of foodstuffs to produce substances with a lower energy content, a process requiring oxygen. The calorific value of fat (the energy released per gram) is twice that of the proteins and carbohydrate. After consuming a meal the metabolic rate increases, the effect being dependent on the class of foodstuffs ingested. Fats and carbohydrates have a smaller effect than protein. This is because the resynthesis of 1 mol ATP during the breakdown of foodstuffs requires more protein energy than that from fat or carbohydrate. If 125 g protein are taken at a meal as meat then metabolism will increase, reaching a maximum after 3–5 h, and then decline slowly. If carbohydrates and fats are eaten in amounts which have an equivalent calorific value, the metabolism increases by only 5–10 per cent. If an ordinary mixed diet is consumed the metabolism is increased by 200–600 kJ (50–150 kcal) daily. The noted effect of protein is termed its specific dynamic action and is 30 per cent of its calorific value. It is in part due to the oxidative deamination of amino acids in the liver and also to a stimulatory effect on cell metabolism of the fatty acid residues remaining after deamination. The stimulating effect of a carbohydrate meal is attributable to the provision of energy to convert a portion of the glucose to glycogen. In reality energy is not being expended, but stored. TTFTF

MCQ 100

Hunger

1. may be increased when certain areas in the hypothalamus are stimulated.
2. is reduced by distension of the gastrointestinal tract.
3. is inhibited by amphetamines.
4. is lost after denervation of the stomach.
5. is closely correlated with blood glucose levels.

Food intake is regulated in the short term, which allows it to adapt to the changing needs brought about by work done, changes in temperature, etc., and in the long term, which allows body weight to be maintained. The hypothalamus is the main centre for integration and relay of information concerning hunger and satiety. Stimulation of the feeding centres located in the medial forebrain bundle at its junction with the paired hypothalamic tracts leads to eating and its destruction to refusal to eat. Stimulation of the satiety centre in the ventromedial nucleus on the other hand causes eating to stop and its destruction leads to extreme obesity. It has been suggested that factors eliciting hunger include contractions of the stomach detected by the mechanoreceptors, reduced glucose availability involving glucoreceptors (glucostatic hypothesis) in the diencephalon and gastrointestional tract, decline in heat production (thermostatic hypothesis) and altered fat metabolism (lipostatic hypothesis). The contractions of the stomach are not a vital part of the sensation of hunger as eating behaviour is little affected by denervation of the stomach or even removal. Eating usually ceases before the absorption of food has corrected the original energy deficit. This results from stimulation of the olfactory receptors, taste receptors and mechanoreceptors in the nose, mouth, throat and oesophagus associated with feeding and possibly also chewing mechanisms. Other factors such as distension of the stomach and stimulation of chemoreceptors in the stomach and upper small intestine may also play a role. In the longer term the changes in metabolism and increased heat production affect receptors in the central nervous system and hence food intake. TTTFF

MCQ 101

In the oral cavity it is possible to

1. distinguish between sweet and sour.
2. assess toxicity.
3. detect low concentrations of metal ions.
4. exert forces of up to a maximum of 15 kg.
5. detect particles down to 10 μm in diameter.

Chewing is a co-ordinated activity involving the teeth of the upper and lower jaw, muscles moving the jaws, the tongue, the cheeks, the palate and floor of the mouth. A crushing force of 14–36 kg and 45–82 kg is exerted by the incisors and molars, respectively. Voluntary mastication is important, but chewing is mainly a reflex activity which produces cutting and grinding of the food. One biological role of taste is to test the edibility of food; toxicity, however, cannot be detected. Taste also influences the process of digestion, the composition of the secretions depending on, for example, whether a sweet or salty taste predominates in the food. Essentially the human can detect four basic tastes: sweet, salt, sour and bitter. Some other tastes are not easily included in this grouping, as for example metallic and alkaline tastes. Sweet tastes are most easily perceived at the tip of the tongue, bitter at the back, sour at the edge and salt both at the tip and the edge. The taste receptors are collected in groups of ten upwards to form taste buds, set in the edge and upper surface of the tongue and walls of the posterior pharynx. Large vallate papillae at the base of the tongue contain 200 taste buds, whereas fungiform and foliate papillae on the anterior and lateral part of the tongue contain relatively few. It is thought that the stimulating chemicals interact with the specific receptor sites on the cell membrane and produce a depolarizing receptor potential. The afferent nerve fibres involved travel in the chorda tympani branch of the facial nerve and glossopharyngeal nerve. Sourness requires the presence of hydrogen ions, the reference being HCl, whose threshold is about 0.1 mmol l^{-1}. Salty tastes are produced by the anions of organic salts, especially the halogens (chlorine, bromine and iodine) and sulphate and nitrate. Bitter taste is given by a variety of substances, often organic. It is believed that sweet-tasting substances bind to receptor membranes non-ionically via hydrogen bonds. Many bitter substances may be converted into sweet-tasting ones by a small change in the molecular structure. TFTTF

MCQ 102

In a healthy adult saliva

1. contains a fixed concentration of sodium ions.
2. provides antiseptic.
3. is secreted within one second of food entering the mouth.
4. secretion is primarily under hormonal control.
5. is essential for the complete digestion of starch.

Saliva has a number of functions, including the initiation of the digestion of starch through the action of amylase. It acts as a solvent for some of the constituents of food and hence allows them to stimulate the taste buds. It helps to keep the mouth moist and thus aids in the swallowing of food and in speech. It also prevents dental caries and, by preventing the pH of the mouth from becoming too acid, prevents decalcification of the teeth. The control of salivary secretion is under nervous control. Stimulation of the parasympathetic nerves produces copious salivary secretion and associated vasodilatation. The latter is produced by several mechanisms, the main one being release by kallikrein of kinins. There is some modification of the composition of saliva by hormones such as aldosterone. Reflex salivation is produced by the smell of food and the presence of food in the mouth and will occur within one second of food entering the mouth. The conditioned reflex described by Pavlov is not important in man. The basal rate of secretion is about 0.5 ml min^{-1}, although it decreases at night. During eating the rate of secretion may increase up to a maximum of 7 ml min^{-1}. The composition of saliva is influenced by the rate of salivary secretion. The primary secretion resembles that of an ultrafiltrate of plasma, but the composition is modified in the ducts, with sodium and chloride being reabsorbed and potassium ions and bicarbonate being secreted. At high rates of flow there is little modification, the composition approaches that of plasma and the osmolality is 220 mOsm kg^{-1}. At low flow rates the secretion is relatively rich in potassium and has an osmolality of 100 mOsm kg^{-1}. Saliva also contains many proteins, including the enzymes amylase and kallikrein as well as glycoproteins. FTTFF

Swallowing in man

1. involves a reflex activity.
2. is solely under the control of the medulla.
3. is associated with elevation of the larynx.
4. demonstrates that not all skeletal muscle is under voluntary control.
5. affects hearing.

It is believed by some that the entire process of swallowing is a reflex phenomenon. Others have suggested that it comprises a voluntary stage in which the bolus is squeezed into the pharynx, a pharyngeal stage involving the involuntary movement of the bolus into the oesophagus and finally an oesophageal phase in which the bolus passes to the stomach. Swallowing is facilitated by the activity of higher centres. It is initiated by the stimulation of many receptors in the mouth and pharynx. Impulses pass via the glossopharyngeal and vagal nerves to the medullary swallowing centre. As the bolus moves into the pharynx the soft palate is elevated and opposes the back wall of the pharynx, sealing the nasal cavity and preventing food from entering it. The swallowing centre also inhibits respiration, raises the larynx and closes the glottis, bringing the vocal cords together and keeping the food from getting into the trachea. As the tongue forces the food back into the pharynx the bolus tilts the epiglottis backward to cover the glottis. This period of swallowing when the bolus passes through the pharynx lasts one second. Peristaltic waves then convey the bolus down the oesophagus to the stomach. Normally a peristaltic wave passes along the oesophagus in ten seconds. Skeletal muscle surrounds the upper third of the oesophagus and smooth muscle the lower third. Thus the swallowing centre must act on the somatic nerves supplying skeletal muscle and the autonomic nerves supplying smooth muscle. The pressure in the mouth is atmospheric and that of the oesophagus in the thoracic cavity is 5–10 mmHg (0.67–1.33 kPa) less than atmospheric as food tends to move into the oesophagus. Regurgitation of the food from the stomach into the oesophagus is prevented by the physiological sphincter. TFTTT

MCQ 104

The electrical complex in the smooth muscle of the distal stomach

1. repeats at a frequency of three per minute.
2. is absent during fasting.
3. always has a plateau phase.
4. sometimes includes bursts of calcium spikes.
5. always induces a contraction.

Gastric motility, as with motility in other areas of the gastrointestinal tract, depends on the co-ordinated contraction of the smooth muscle layers. There are three muscle layers in the stomach: the outer longitudinal, middle circular and inner oblique. Gastric motility is dependent on the myogenic properties of the smooth muscle layers themselves, on the activity of the intramural plexuses and the extrinsic nerves, principally the vagus, and on the gastrointestinal hormones, including gastrin and secretin. The patterns of motility differ in the proximal and distal regions of the stomach. The proximal stomach receives the food and acts to store it with minimal mixing activity. The dominant form here is gentle and is called tone contraction. These contractions help to deliver ingested food to the distal stomach. Acid is mixed in the distal stomach, where there are strong peristaltic contractions. The frequency of the peristaltic contractions depends on the rhythmic depolarization of the resting membrane potential. There is rapid depolarization followed by a partial repolarization and then a plateau phase, after which there is complete repolarization. The plateau phase is quite variable and sometimes is not seen, and then no contraction occurs. Sometimes calcium action potentials are superimposed on the plateau. The rhythmic electrical activity occurs at a frequency of three cycles per minute. The electrical rhythm passes down the stomach, the cells with the highest frequency of spontaneous activity establishing the frequency for the whole stomach. The frequency of peristalsis is constant, although it can be increased by gastrin. TFFTF

MCQ 105

The secretion of hydrochloric acid

1. is from the chief cells.
2. occurs in the fundus and body of the stomach.
3. is initially at pH 4.
4. has an initial chloride concentration of approximately 150 mM.
5. produces excess hydroxyl ions in the plasma, resulting in the alkaline tide.

In the fundus and corpus the columnar epithelial cells lining the stomach extend into the gastric pits into which three to seven gastric glands open. Three different types of cells are seen in these glands. At the neck of the gland are the neck cells secreting mucus. In the base of the gland are the most numerous cells, the chief or peptic cells, which secrete pepsinogens, inactive precursors of proteolytic enzymes. The third type of cell, lying largely in the middle gland, are the oxyntic or parietal cells which produce hydrochloric acid. This helps in the breakdown of connective tissue and muscle fibres, activates pepsin and provides an environment of low pH in which the enzyme can act. The human stomach secretes about 2 l HCl per day, with a pH of 0.8–1.5. The mechanism of gastric acid secretion is not clearly defined. It is known that hydrogen and chloride ions are actively pumped from the plasma into the secretory canaliculus. The chloride is derived from the plasma, whereas hydrogen ions are formed within the parietal cells, probably being produced by the dissociation of water. For each hydrogen ion excreted a hydroxyl ion is left in the cell and this is neutralized by a further hydrogen ion derived from carbonic acid, formed from carbon dioxide under the influence of carbonic anhydrase; the bicarbonate also produced diffuses back into the plasma, resulting in the alkaline tide. The contribution of carbonic anhydrase to acid secretion means that inhibitors of its action suppress acid secretion. FTFTF

MCQ 106

The value of the stomach includes the ability to

1. act as a reservoir.
2. sterilize ingested food by reducing the pH.
3. homogenize the dietary intake.
4. synthesize and secrete intrinsic factor.
5. contribute to the enzymic digestion of all classes of foodstuffs.

While the stomach is not essential for life, it has a number of important functions. One is the storage of food, the upper part of the stomach being able to hold quite large quantities of food without an increase in pressure. This is due to the properties of the smooth muscle, the arrangement of the fibres and to receptive relaxation, a reflex involving the vagus. The bolus of food is reduced to a semi-fluid mass, the chyme, in the stomach by the rhythmic contractions of the stomach and by chemical activity. Protein digestion is initiated in the stomach through the action of pepsin and hydrochloric acid, which provides an optimum pH for the enzyme's activity and also denatures the protein. The acid medium also serves to kill ingested bacteria. The stomach releases food into the duodenum at a controlled rate. Another significant function of the stomach is the production of intrinsic factor. This is a glycoprotein of molecular weight 60 000 which is necessary for the absorption of vitamin B_{12}. As vitamin B_{12} is a complex cobalt-containing vitamin, it would not otherwise be absorbed. Vitamin B_{12} or cyanocobalamin is necessary for normal erythropoiesis and lack results in anaemia characterized by the presence in the blood of large primitive red cell precursors called megaloblasts (see MCQ 12). This type of anaemia rarely occurs as a result of dietary deficiency, but is due to inadequate absorption resulting either from deficiency of intrinsic factor, as occurs following gastrectomy, or from diseases in which intestinal absorption as a whole is impaired, as for example in coeliac disease (gluten-induced enteropathy), in which the villi are deformed and hence the absorptive surface reduced. TTTTF

MCQ 107

Gastric secretion occurs in response to

1. the presence of food in the mouth.
2. protein products in the intestine.
3. a high pH in the pyloric antrum.
4. administration of atropine.
5. administration of histamine.

Many factors influence the secretion of gastric juice, some of them stimulatory, some inhibitory. Classically gastric secretion has been divided into three phases: cephalic, gastric and intestinal. The stimuli which contribute to the cephalic phase are the sight, smell and taste of food, as well as the process of chewing. The pathway to the parietal cells secreting hydrogen ions involves the vagus, via the nerve plexuses, and gastrin. The contribution of the cephalic phase has been studied using the technique of sham feeding pioneered by Pavlov, in which the ingested food was removed via an oesophageal fistula instead of passing into the stomach. Once the food does reach the stomach, the distension of the stomach, the fall in hydrogen ion concentration (due to the buffering action of the proteins) and the presence of peptides all lead to an increase in hydrogen ion secretion. The mechanisms include long and short neural reflexes, enhanced gastrin release and the direct action of the peptides. The contribution of both nervous and hormonal systems has been established using different types of gastric pouch. There is the Heidenhain pouch, produced from the mid-portion of the greater curvature with the vagal branches severed. Second is the Pavlov pouch, prepared from the same region of the stomach, but with the vagal innervation preserved. Finally there is the Farrel and Ivy pouch, a more completely denervated pouch which is transplanted into the mammary gland region of the abdominal wall. Lastly, there is the intestinal phase of digestion. Some 10 per cent of the acid secreted is accounted for by this phase, which results from a hormonal response. However, distension of the small intestine, the presence of nutrients, as well as an increase in hydrogen ion concentrations and osmolality, all suppress hydrochloric acid release via nervous and hormonal mechanisms. TTTFT

MCQ 108

The rate at which a liquid meal leaves the stomach is slowed by

1. increasing the size of the meal.
2. vagal activity.
3. the presence of triglycerides and the products of their digestion.
4. a 5 per cent glucose meal.
5. 'enterogastrone'.

When a peristaltic wave approaches the larger muscle mass surrounding the antrum it moves more rapidly and produces a more powerful contraction which mixes the contents of the stomach and expels them into the duodenum. There is between the terminal antrum and the duodenum a ring of smooth muscle and connective tissue, the pyloric sphincter. In some animals, such as the dog, no anatomical sphincter can be identified and the gastroduodenal junction may act as a physiological sphincter. The sphincter itself only exerts a slight pressure and is open most of the time. However, when a strong peristalsis arrives at the antrum it also closes the pyloric sphincter so that most of the antral contents pass back into the body of the stomach. The rate at which the gastric contents pass into the duodenum depends on their chemical and physical characteristics. Liquids pass into the duodenum more quickly than solids. Emptying is also influenced by the volume of the stomach contents, the greater it is the more rapid the emptying. Thus the decrease of the amount of a meal remaining in the stomach is semi-exponential. Gastric emptying is slowed by the presence of fats and their digestive products and of acid in the upper intestine, also by hypertonic solutions. The 5 per cent glucose meal is isotonic and empties at the maximum rate. Thus many substances present in the duodenum inhibit gastric emptying and it is envisaged that there are several different receptors, including one which detects acid and one which responds to changes in osmolality. In some cases sectioning the vagi alters the response and the term enterogastric reflex has been used to describe the events. Vagotomy may temporarily slow gastric emptying in man, but the long-term effect is to increase the rate in some or to leave it unchanged in others. The inhibitory response following, for example, ingestion of fat is said to involve hormonal mechanisms and the term enterogastrone has been used to describe the inhibitor(s) released, but it is not yet known if this is a combination of known hormones or an as yet unidentified substance. FFTFT

MCQ 109

The symptoms of the dumping syndrome following gastric surgery may include

1. hyperglycaemia immediately after a meal.
2. a fall in insulin concentrations.
3. weakness and sweating.
4. a fall in plasma volume.
5. a reduction in cardiac output.

If a partial or even total gastrectomy is performed, there is no effect on protein digestion, which proceeds normally in the absence of pepsin. However, the acid normally released by the stomach promotes iron absorption and the intrinsic factor promotes vitamin B_{12} absorption, so that anaemia develops. Such patients also suffer from the dumping syndrome, which comprises attacks of nausea, vomiting and diarrhoea, with dizziness and pallor followed by flushing, tachycardia, sweating and weakness following a meal. A number of factors contribute to the syndrome. The hyperglycaemia consequent on the rapid emptying of the meal and rapid absorption of glucose from the intestine results in an abrupt rise in insulin, which in turn produces hypoglycaemia. This fall in blood glucose is largely responsible for the late postprandial symptoms, such as weakness, dizziness and sweating. The rapid entry of hypertonic meals into the intestine stimulates movement of water into the intestine and a consequent reduction in plasma volume sufficient to reduce the cardiac output. If after a standard glucose meal the fall in plasma volume is less than 8 per cent, there are no dumping symptoms, while patients with a fall in plasma volume greater than 17 per cent always complain of such symptoms. Patients with a clinical syndrome always have a fall in plasma volume greater than 9 per cent. TFTTT

MCQ 110

Chyme is mixed or moved along the small intestine

1. by controlled tonus changes.
2. by segmentation contractions depending on the myenteric plexus.
3. exclusively by peristalsis.
4. a process unaffected by section of the vagus.
5. through the inherent properties of the smooth muscle.

The movements of the small intestine are necessary to mix the chyme from the stomach with digestive secretions from the pancreas, liver and intestine itself. They also bring the products of digestion into contact with the absorbing cells and propel the luminal contents to the ileocaecal junction. The mixing is brought about by segmenting contractions, while the peristaltic contractions bring about propulsion over short distances. The segmenting contractions, which are dependent on the circular muscle, are localized, involving only 2 cm of the intestine. The rate of segmentation is independent of extrinsic nerves. Each region of the intestine when separated from other regions has a unique slow wave activity, the frequency of which decreases distally. The peristaltic contractions, which depend more on the longitudinal muscle, produce a wave moving slowly at a rate of 1–2 cm s^{-1} down the intestine. The contractions of the small intestine are modified by a number of factors including inherent smooth muscle cell activity, the action of intrinsic and extrinsic nerves and circulating or locally released chemicals. There are two main nerve plexuses in the small intestine: the myenteric or Auerbach's plexus, which innervates both longitudinal and circular muscle; and the submucosal or Meissner's plexus. The axons from sympathetic fibres may synapse either with the myenteric plexus or directly with the smooth muscle. Stimulation of the sympathetic nerves usually inhibits contractions. In addition to an effect on the myenteric plexus the response may be brought about by vasoconstriction and diffusion of neurotransmitters from blood to muscle cells and also general sympathetic stimulation of adrenaline and noradrenaline from the adrenal medulla. Gastrointestinal hormones influence motility. For example, gastrin and cholecystokinin/pancreozymin (CCK/PZ) stimulate motility, while pancreatic glucagon inhibits muscle activity. FTFTF

Pancreatic secretion

1. contains bicarbonate.
2. contains the enzyme trypsin.
3. has no afferent nerve control.
4. is under the influence of cholecystokinin.
5. is affected by secretin.

The pancreatic secretions contain two major components—an alkaline-rich fluid and an enzyme-rich fluid—the two substances being secreted into the same ducts. The alkaline fluid, which is isotonic with plasma, helps neutralize the acid entering the duodenum. It is secreted by the cells lining the early portion of the ducts leading from the acinar cells. The main cations are Na^+ and K^+ and are present in similar concentrations to those in the plasma. The major anions are HCO_3^- and Cl^-, the concentrations as in saliva depending on the rate of secretion, but in this instance bicarbonate increases with increasing flow and chloride concentrations decrease, because there is little time for passive Cl^- and HCO_3^- exchange. The active process involved in secretion is transport of sodium from the plasma to the lumen. Sodium transport is also linked to exchange of potassium and hydrogen ions from carbonic acid. The enzyme-rich secretion is produced by the acinar cells. It contains enzymes capable of hydrolysing all major classes of nutrients. Many are secreted in the inactive form, as for example the proteolytic enzyme trypsin, which is secreted as trypsinogen, and chymotrypsin, which is secreted as chymotrypsinogen. Trypsinogen spontaneously changes into the active form, but the process is activated by the proteolytic enzyme enterokinase and by activated trypsin itself. Another enzyme with protein as its substrate is carboxypeptidase, which splits off the terminal amino acid with the free carbonyl group. Several enzymes in the pancreatic juice are concerned with the breakdown of lipids, a key one being lipase. There is also present an enzyme pancreatic amylase, which like that from the saliva is an α-amylase, hydrolysing 1,4-glucosidic bonds. Other enzymes include ribonuclease and deoxyribonuclease, which split nucleic acids into free mononucleotides. As with gastric secretion, release of pancreatic juice may be divided into three phases: cephalic, gastric and intestinal. The response involves neural activity via the vagus and the hormones gastrin, secretin and CCK/PZ. TFFTT

MCQ 112

Bile

1. is formed in the gall-bladder.
2. is released through the action of secretin.
3. is the main excretory pathway for urea.
4. contains cholesterol.
5. pigments are converted to urobilinogen by intestinal flora.

Bile is formed in the liver, the daily output being 700–1200 ml. The chief constituents are bile acids, phospholipids (particularly lecithin), cholesterol, bile pigments and electrolytes. The bile acids form large aggregates, so that the osmotic activity is much lower than would be expected. Bile acids, like cholesterol and phospholipids, are involved in solubilization of fat in the small intestine. They are water-soluble derivatives of cholesterol. Two primary bile acids, lithocholic acid and deoxycholic acid, are formed in the intestine by bacterial action. The major bile pigment, bilirubin, is an intermediary in the elimination of haem, which is produced in the breakdown of haemoglobin as described in MCQ 12. The conversion to bilirubin takes place in the spleen and bone marrow and also to a limited extent in the liver. The cells of the liver extract bilirubin from the circulation and secrete it into the bile by an active process. Other minor sources of bile pigments are myoglobin and cytochromes. Bile pigments are yellow, but are modified in the intestine to form the pigments which give the faeces their colour. Some pigments are reabsorbed from the intestine and are excreted in the urine, giving it a yellow colour. If the bile duct is blocked the pigment spills over into the circulation, so that the patient looks jaundiced. The faeces are pale and the urine very dark. The bile duct may be blocked by cholesterol or bicarbonate stones. They can calcify because the concentrations of calcium in the gall-bladder can become quite high as water is reabsorbed during the storage of bile. Bile secretion is stimulated by agents termed choluretics, which include bile acids, and can be modified by gastrin, CCK/PZ and vagal mechanisms. Contraction of the gall-bladder and relaxation of the sphincter of Oddi at the entry of the bile duct to the intestine is stimulated by CCK/PZ. FFFTT

Absorption of dietary fats

1. mainly occurs in the terminal ileum.
2. depends on the complete hydrolysis of fats to fatty acid and glycerol.
3. if decreased, reduces absorption of vitamins A, D and K.
4. takes place only in the lymphatics.
5. is dependent on the presence of calcium ions.

In Western Europe the diet contains some 60–100 g fats, which since they are insoluble in water present some physicochemical problems in digestion. Initially the mechanical mixing in the stomach and upper small intestine breaks fat droplets into minute particles, at the surface of which the water-soluble lipase can act. Emulsifying agents such as bile salts, cholesterol and fatty acids prevent the droplets from coalescing. The products of the action of pancreatic lipase—2-monoglycerides and free fatty acids—are not really soluble in water but maintain the dissolved form in small stable units 5 μm in diameter, called micelles. Their formation depends on the bile acids and they also contain lecithin and the fat-soluble vitamins A, D and K. Fatty acids and monoglyceride micelles may be present in concentrations 1000 times greater than those of the individual molecules, so that even though the diffusion coefficient of micelles through the unstirred layer adjacent to the epithelial cell is less than that of fatty acids, allowing presentation of greater amounts of products of fat digestion to the absorptive surface, the total bile pool is insufficient for the digestion of a meal, so that the bile acids are recycled, being reabsorbed from the terminal ileum. This is distal to the site of absorption of fatty acids and monoglycerides, which are absorbed mainly in the proximal jejunum. They are then resynthesized into triglycerides which are enveloped in a coating of protein, cholesterol and phospholipid. These particles, the chylomicrons, are released from the intestinal mucosa cells by exocytosis, pass into the extracellular space and thence to the lymphatics. They then enter the blood via the thoracic duct. Lipoproteins also pass into the lymph as very low density lipoproteins. FFTFT

MCQ 114

The absorption of water from the gastrointestinal tract

1. is greatest in the ileum.
2. is mainly a passive process.
3. can occur uphill in the ileum.
4. can be modified by aldosterone.
5. in response to an osmotic gradient is via the paracellular pathway.

Apart from the water ingested, 8–9 l secretions enter the upper alimentary canal daily. Of this, 1 l enters the large intestine, which normally absorbs all but 100 ml, which passes out in the faeces. The apical and basolateral membranes of the epithelial cells of the small intestine are permeable to water, so that an osmotic gradient is set up across the epithelium. The resulting passive movement of water is both transcellular and paracellular. Sodium and chloride are transported into the lateral spaces, increasing the osmolality there and drawing water along by osmosis. As water enters the lateral intracellular spaces, they swell and the hydrostatic pressure increases. Since resistance to hydrodynamic flow is high at the apical end and low at the basal end and water tends to move towards the blood, absorption of water in the stomach is negligible. The duodenum, jejunum and ileum all absorb water, the ileal epithelium being able to transport water uphill. This is not by active pumping of water, but occurs through the osmotic coupling of water flow to ion transport. Increased aldosterone secretion, as may occur in sodium deficiency, leads to increased reabsorption of sodium in the colon and water follows passively. If the luminal contents have a high osmolality then water enters the lumen. This occurs in diarrhoea produced by non-absorbable purgatives and purgatives such as phenolphthalein, which block glucose reabsorption. As mentioned in MCQ 109 it is also seen in the dumping syndrome. FTTTF

MCQ 115

The intestinal absorption of

1. glucose occurs by passive diffusion.
2. xylose (a pentose) is faster than glucose (a hexose).
3. glucose is inhibited by the presence of sodium ions in the intestinal lumen.
4. amino acids may be studied by the everted sac technique.
5. peptides into the enterocytes involves specific carriers.

Carbohydrates are absorbed from the small intestine in the form of monosaccharides. The breakdown of carbohydrates is initiated by salivary amylase, continued by the pancreatic enzymes and completed in the intestine under the influence of oligosaccharides associated with the epithelial cell brush border. Special carrier mechanisms exist to transfer monosaccharides from the lumen. The rate of transfer of individual sugars varies. For example glucose and galactose are transferred more rapidly than the others. The glucose and galactose transfer mechanisms are energy-dependent, sodium-dependent and unequally distributed along the intestine. A carrier is believed to be involved, which is mobile and combines with glucose and sodium, transferring them across the barrier. Glucose and sodium are released at the inner face and the sodium is transported out. Thus the energy for glucose transport is indirectly provided by the transport of sodium. The products of protein digestion are usually absorbed in the form of amino acids, although some peptides and proteins are absorbed. Amino acids and peptides are absorbed via sodium-dependent carrier mechanisms. Protein movement probably occurs via extrusion zones at the tips of the villi or by remnants of mechanisms present in the fetus and newborn. The small intestine of the newborn displays a transient permeability to some macromolecules, which is probably one way in which passive immunity can be conferred. Hydrolysis of proteins begins in the stomach under the influence of pepsin, the action of which is terminated when the gastric contents reach the more alkaline duodenum. Here pancreatic secretions contribute to breakdown. As well as hydrolysis in the lumen, the process can occur at the brush border or within the cytoplasm. Once the peptides are broken down to individual amino acids, there are at least four amino acid transfer systems to aid in their absorption: one concerned with the movement of neutral amino acids, others for dibasic and dicarboxylic amino acids and finally one for imino acids and glycine. Transport may be studied *in vivo* by following the appearance of substances in the bloodstream, by sampling from the lumen via tubes, including triple tubes which allow balloons to be inflated to isolate a portion of the intestine and by surgical preparation of loops of intestine. Several *in vitro* techniques have been developed, the most commonly used being the everted sac technique. FFFTT

MCQ 116

The liver is important in

1. secretion of hormones.
2. synthesis of vitamins.
3. protein synthesis.
4. carbohydrate metabolism.
5. fat metabolism.

The functions of the liver are numerous. but may be summarized as follows. The liver acts as a storage organ, has a detoxifying function and is involved in synthesis and metabolism, bile secretion and formation and destruction of red cells. The liver plays a central role in intermediary metabolism. Carbohydrates are stored here in the form of glycogen and glucose is synthesized from non-carbohydrate sources (gluconeogenesis). The liver is also the only site of glucose-6-phosphatase action, the enzyme required for the breakdown of glycogen. When fats are used in the body they are withdrawn from adipose tissue and pass to the liver. Normally the fat is broken down as it arrives, but under certain circumstances—as on a high fat diet or during starvation (when there is a lack of lipoprotein)—the neutral fat content of the liver may be increased. In circumstances of carbohydrate deficiency metabolism of liver fat may play a role in the maintenance of blood sugar and this is associated with the production of ketone bodies, the liver being the only organ which produces them on any significant scale. In common with many other tissues, liver deaminates surplus amino acids with the production of urea. It also is an important site for protein synthesis, namely plasma proteins, fibrinogen and thrombin. It is not, however, the site of synthesis of vitamins; rather it stores them, e.g. vitamins A, D and B_{12}. In the fetus the liver is the site of formation of red cells and throughout life it removes bilirubin from the blood following destruction of red cells. Liver is involved in the breakdown of many substances. These may be substances such as hormones, including vasopressin and aldosterone, or toxic substances, in which case the liver is exerting a protective action. Detoxification may occur by conjugation with, for example, sulphate, glycine, glucuronic acid or acetic acid. Many drugs and hormones containing hydroxyl groups whether alcoholic or phenolic combine with glucuronic acid. Urinary steroids are excreted in the form of glucuronides. The liver may also bring about complete destruction of a substance. Many compounds foreign to the body are destroyed in the liver by complete oxidation as, for example, short-acting barbiturates. FFTTT

Defecation

1. is an important mechanism for the elimination of toxins from the body.
2. is controlled by the conscious relaxation of the internal anal sphincter.
3. occurs on stimulation of the nervi erigentes.
4. may be aided by the Valsalva manoeuvre.
5. requires the integrity of the sympathetic nerves supplying the rectum.

The functions of the large intestine are to store material received from the small intestine, to absorb water and salts, to allow some secretion and excretion and to house bacteria which produce substances such as vitamin K. The intestinal flora also break down carbohydrates, by a process of fermentation, and proteins yielding poisonous amines, including indole, and potentially toxic substances such as histamine and tyramine. The faeces normally contain 65–80 per cent water. The solid matter comprises about one-third bacteria which have grown in the digestive tract, undigested food residues such as cellulose and keratin, some inorganic material, as for example insoluble and iron salts, some fat and finally desquamated epithelial cells. The colour of the faeces comes from the modified bile pigments. In the newborn infant defecation is a reflex event stimulated by increased pressure in the rectum. Impulses travel via visceral afferent fibres to the autonomic centres in the sacral part of the spinal cord. The parasympathetic fibres in the nervi erigentes are stimulated so that the smooth muscle of the colon, especially the longitudinal muscle, is stimulated and the sphincter muscles of the anus relaxed. This results in the propulsion of faecal material through the colon into the rectum and out through the anus. However, during the first few years of life tone develops in the external sphincter and reflex defecation is brought under conscious control. Defecation then takes place when relaxation of the external sphincter occurs. Defecation is aided by an increase in the intra-abdominal pressure brought about by increased tension in the muscles of the abdominal wall and contraction of the respiratory muscles to produce a lowering of the diaphragm. A greater rise in intra-abdominal and intra-thoracic pressure can be brought about by the Valsalva manoeuvre, in which expiration takes place against a closed glottis. At the same time the pelvic floor is pulled upwards and outwards on the anus to evaginate the faeces downwards. TFTTF

MCQ 118

Characteristic of all hormones is that

1. their secretion is independent of the central nervous system.
2. they enter the systemic circulation.
3. they bind to specific receptors in proportion to their plasma concentration.
4. they are produced only in the classical endocrine organs.
5. they have specific actions.

Classically a hormone has been defined as a substance produced in a ductless or endocrine gland and secreted into the systemic circulation to act at a second organ. This has not been found to be true of all hormones. For example the hypothalamic releasing hormones are secreted into the hypothalamo-hypophyseal portal system. Furthermore not all hormones have a classical endocrine function. Some are paracrines producing an effect on the surrounding cells and not passing into the circulation, while others have a neurocrine function, acting as neuromodulators. Finally some are autocrine. Thus not all hormones are produced in endocrine glands: some are produced in nerve cells, including hypothalamic neurones, while others are produced in the placenta, gastrointestinal tract and kidney. Traditionally an approach to studying the possible endocrine function of an organ is to monitor the effect of an extract of the tissue. The response to removal of the organ and the effect of reimplantation can be followed. Many endocrine organs function normally on transplantation as they do not have a nervous supply. However, the pituitary and the adrenal glands are under the control of the nervous system. Hormones form part of the control system of the body. They have specific effects on their target organs, which are dependent on their binding to specific receptors. There is, however, a limited number of these sites so that their binding cannot increase indefinitely. Some hormones such as growth hormone, thyroxine and insulin have a widespread activity throughout the body, while other hormones such as adrenocorticotrophic hormone (ACTH) and thyroid-stimulating hormone (TSH) act largely on a single tissue. Hormone actions may be classified in many ways, but three headings may be employed: those hormones which control growth and sexual development, those which bring about physiological adjustments to the changing environment and those which keep physiological parameters constant. FFFFT

MCQ 119

Among the steroid hormones may be included

1. cortisol.
2. vitamin D.
3. aldosterone.
4. corticotrophin-releasing factor.
5. adrenocorticotrophin.

Hormones classified according to their chemical structure fall into three main groups. First are the protein and peptide hormones which include molecules as small as thyrotrophin-releasing hormone (TRH) comprising three amino acids and molecules as large as the glycoprotein hormones of the anterior pituitary or growth hormone, with 190 amino acids. Second are the steroid hormones which are derivatives of cholesterol and are produced by the adrenal cortex and the gonads. Vitamin D may also be classified as a hormone as vitamin D_3 (cholecalciferol) is formed in the skin and then metabolized in the liver and kidney, giving the active derivative 1,25-dihydroxycholecalciferol which then acts on target organs, reaching them via the blood stream. The final group comprises the derivatives of tyrosine, the monoamines dopamine, adrenaline and noradrenaline and the thyroid hormones triiodothyronine (T_3) and thyroxine (T_4). Protein and peptide hormones are synthesized using normal protein synthetic mechanisms. Peptide hormones are generally formed as pre-prohormones. The pre fragment is cleaved as the peptide leaves the ribosome. The protein then travels through the endoplasmic reticulum and some carbohydrate moieties may be added at this stage. Further carbohydrate moieties may be added on transfer of protein to the Golgi complex. The complete but still inactive protein is then packed into vesicles containing specific endopeptidases, which are activated on secretion, allowing the active hormone to be released on exocytosis. In the synthesis of steroid hormones the first step is the hydrolysis of cholesterol esters found in the fat droplets, which are characteristic of steroid-producing cells. The cholesterol is then transported to the mitochondria where side-chain cleavage results in the production of pregnenolone, which is transported out of the mitochondria. Further transformations occur on smooth endoplasmic reticulum or in the mitochondria. TTTFF

MCQ 120

Endocrine control systems may

1. involve negative feedback mechanisms.
2. have both chemical and nervous inputs.
3. involve polypeptide-releasing factors.
4. involve positive feedback mechanisms.
5. involve a series of hormonal intermediates.

Some hormones are secreted at a relatively constant rate, while others show variation in response to physical or emotional stimuli or can show cyclical variations which can be of a few minutes', hours' or days' duration. The elements of control systems have been discussed in MCQ 5. The simplest form of hormonal control system is a negative feedback system operating to maintain a given physiological parameter within fixed limits. This is exemplified by glucose or calcium homeostasis. Thus an increase in blood glucose results in the release of insulin, which in turn lowers blood glucose. Equally if the plasma calcium concentrations fall, parathyroid hormone (PTH) secretion is stimulated, which leads to elevation of blood calcium. A more complicated system is seen in the control of the anterior pituitary hormones and the target glands they control. In these systems both open-loop and closed feedback-loop systems exist. Anterior pituitary hormone secretion is controlled by hypothalamic polypeptide-releasing factors. The anterior pituitary trophic hormones, which act on a number of endocrine glands, the thyroid, adrenals and gonads act to stimulate hormone release. The secretions released from the target organs then exert a negative feedback on the anterior pituitary or hypothalamus. It is also possible that short loops exist, whereby anterior pituitary secretions feed back on the hypothalamus. Hypothalamic factors such as corticotrophin-releasing factor are released in a pulsatile fashion, so that ACTH and cortisol show a similar pattern. Additionally a circadian rhythm is seen, with plasma concentrations of ACTH and cortisol being highest in the morning, falling to a nadir at midnight. In some situations positive feedback systems exist. Positive feedback of oestrogen is responsible for the luteinizing hormone surge at the time of ovulation. Positive feedback is also believed to operate in the control of oxytocin release during parturition. During delivery stimulation of the cervix results in the release of oxytocin, which increases uterine activity and causes further stimulation. TTTTT

MCQ 121

In the life cycle of a hormone

1. peptide hormones may be synthesized in the form of a prohormone.
2. transport in the plasma is usually in the bound form.
3. the plasma concentration depends almost entirely on the secretion rate.
4. the products of its breakdown may be excreted in the urine.
5. conversion to an active form may occur in the tissues.

As indicated in MCQ 119 peptide hormones are synthesized in the form of prohormones. For example insulin is formed as proinsulin, which is cleaved to yield insulin and C peptide. Once released from the gland, many hormones travel in the bound form. Specific binding hormones such as thyroxine-binding globulin and corticosteroid-binding globulin exist, although some binding to albumin also occurs. Plasma binding of hormones helps prevent their loss in the kidney. Binding proteins also influence the concentrations of free hormone, which is the active form of the hormone. Care must be taken, therefore, in interpreting the meaning of total plasma concentrations of a hormone as they may merely reflect elevation of binding protein, as may be seen for example on ingestion of oral contraceptives. Plasma concentrations thus do not just depend on the rate of hormone secretion, but also on concentrations of plasma-binding proteins and the rate of clearance. The metabolic clearance rate is analogous to renal clearance and represents the sum of all the individual clearance rates of the different organs. The main organs for clearance are the liver and the kidney and the products of breakdown may be excreted in the urine, which allows the determination of the clearance rates of certain steroids. If a steroid is excreted in the urine as a unique metabolite, the amount by which an administered radioactively labelled steroid is diluted by the endogenous hormone can be estimated from the specific activity of the metabolite in urine. Alternatively, the hormone can be infused at a given rate and the equilibrium concentration and the rate of disappearance from plasma determined by estimating the plasma concentrations. Little of the excreted hormone is inactivated in the target organ. However, conversion to the active forms of the hormone may occur: thyroxine is converted to the more potent triiodothyronine (T_3). It is suggested that if sufficient T_3 is present then the metabolically inactive reverse T_3 is formed instead. The action of testosterone at most target sites requires its further metabolism to 5α-dihydrotestosterone. TFFTT

MCQ 122

Hormones can function by

1. binding with a cell membrane receptor.
2. entering the cytoplasm.
3. causing the release of an intracellular messenger.
4. controlling membrane permeability.
5. a permissive action.

The mechanism of action of a given hormone depends on its chemical structure. Protein and peptide hormones bind to receptors located in the plasma membrane and produce their effects via a second messenger, which is generally cyclic AMP or Ca^{2+}, the possible exceptions being insulin, growth hormone and prolactin. Binding to the receptor is followed by increased binding of GTP to the associated nucleotide regulatory protein, which in turn causes activation of the catalytic subunit for the conversion of ATP to cyclic AMP. Binding of cyclic AMP to the regulatory subunit of inactive protein kinase results in dissociation of the complex and activation of the catalytic subunit. Protein kinases have important effects on protein synthesis and are involved in the phosphorylation of nuclear histones and perhaps other nuclear proteins. Consequently RNA synthesis and protein synthesis may be affected. Hormones may also influence the permeability to a number of substances such as glucose and amino acids. Calcium ions and phospholipid turnover may also contribute to hormone action. In contrast to the cyclic AMP system, where generation of cyclic AMP is nearly always due to direct coupling between hormone and receptors, the change in the calcium signal may not be the initial consequence of hormone receptor interaction. Cytosolic calcium concentrations depends on a number of mechanisms, including breakdown of phosphatidyl inositol. An increase in calcium concentration in the cytosol leads to activation of calcium-dependent enzymes which involve calmodulin as an intermediate. Steroid and thyroid hormones act differently, entering the cell. Steroid hormones bind with receptors in the nucleus and alter the activity of the DNA-dependent RNA polymerase, with a subsequent change in the rate of production of specific proteins. However, cortisol sometimes has a permissive action, not having a direct effect, but allowing other reactions to proceed. Thyroid hormones enter the nucleus and also stimulate DNA-dependent RNA polymerase, which translates RNA from a DNA template. There is a switch to the formation of catabolic enzymes. TTTTT

MCQ 123

Cyclic AMP (adenosine-3′,5′-monophosphate)

1. is formed from ATP.
2. mediates the effect of steroid hormones.
3. concentrations in responsive cells increase prior to the physiological response to a given hormone.
4. effects are mediated through the action of protein kinase.
5. actions are inhibited by theophylline.

Cyclic AMP acts as a second messenger in the response of many peptide hormones but not of steroid hormones, as indicated in MCQ 121. It is formed from ATP under the action of adenylate cyclase. It has been suggested that a number of criteria should be met before it can be considered that cyclic AMP is a mediator of the response in the system under study. These are:

(a) that the hormone should cause an increase in cyclic AMP concentrations;
(b) that the increase in cyclic AMP should precede the physiological response:
(c) that the hormone should activate adenylate cyclase in broken cell preparations;
(d) that inhibitors of phosphodiesterase, an enzyme which breaks down cyclic AMP, should potentiate the effect of the hormone;
(e) that exogenous cyclic AMP or an analogue should mimic the action of the hormone;
(f) that cyclic AMP-dependent protein kinases are present in the cell and their activation is seen with hormonal stimulation.

Protein kinases catalyse the transfer of phosphate from ATP to a protein acceptor. Cyclic AMP-dependent protein kinases are widely distributed and many proteins, including phosphorylase kinase, have been shown to be substrates. Another cyclic nucleotide, cyclic GMP, has been demonstrated which may have a second messenger function, but its full biological role has to be established. It was believed that cyclic AMP was the only important second messenger involved in the action of protein and peptide hormones. It is now known that other systems are involved, including the calcium system and the related phospholipid-activated calcium-dependent kinase system. Phosphatidyl bisphosphate in the membrane is split through the action of phospholipase C to yield inositol trisphosphate and diacyl glycerol. The former acts to mobilize intracellular calcium stores and the latter to activate protein kinase C, which like cyclic AMP phosphorylates proteins and enzymes, leading to different cellular actions. TFTTF

MCQ 124

The advantages of the technique of radioimmunoassay over that of bioassay include

1. its relative simplicity.
2. the ability to process large numbers of samples.
3. its invariably greater specificity.
4. its greater sensitivity on all occasions.
5. its precision.

Measurement of hormones is difficult as they circulate in very low concentrations (as low as picomoles per litre or one million millionth of a mole). In order to perform estimations use is made of the unique properties of biological systems to recognize different molecules. The receptor sites employed in the bioassay are the tissue receptors and, in the immunoassay, antibody. Bioassay entails the determination of the potency of a hormone by noting its biological effect as compared with a standard preparation of defined activity. This type of assay is not routinely used nowadays, but is important in the standardization of hormone preparations. Hormones are defined in terms of their activity and it is only bioassay which gives a measure of biological activity. Bioassays, however, are difficult to perform and have only a very limited throughput. Routine assays may be performed using radioimmunoassay. This is one of a group of assays termed binding assays, which depend on the progressive saturation of a specific binder by the substance being quantified and entails the subsequent determination of the proportion which is not bound. The most common classification of binding assays depends on the binder used, radioimmunoassay employing antibodies, competitive protein binding assays employing a protein binder and receptor assays employing the naturally occurring cell receptor. The distribution of ligand between the bound and free form may be determined using physicochemical separation, the distribution being followed by incorporation of a tracer consisting of a small amount of labelled ligand (usually radioactive). If fixed amounts of antibody and labelled antigen are allowed to react, then at equilibrium they will form an antibody–antigen complex, a proportion of antibody and antigen remaining free. If the antibody concentration is held constant and the amount of antigen increased by addition of unlabelled material, then at equilibrium more of the antigen (labelled and unlabelled) will be free. The concentration of the ligand in an unknown sample can then be calculated from a standard curve, where the distribution between bound and free ligand is taken for a set of standards prepared with the same fixed concentration of labelled ligand and antibody. TTFFT

MCQ 125

Hormones secreted by the basophils of the anterior pituitary include

1. growth hormone.
2. prolactin.
3. adrenocorticotrophin (ACTH).
4. thyrotrophin-releasing hormone (TRH).
5. follicle-stimulating hormone (FSH).

The anterior pituitary produces six main hormones: four trophic hormones which act on other endocrine glands—ACTH, TSH, FSH and luteinizing hormone (LH)—and two hormones with widespread actions—growth hormone and prolactin. The cells producing hormones are arranged in clumps or cords separated by the capillaries arising from the hypophyseal portal vessels. The cells are named according to the hormones they produce: thyrotrophs, cortitrophs, etc. An earlier classification was based on the staining characteristics of the cells, the cells being divided into chromophobes (poorly staining) and chromophils (well stained). The latter may be further subdivided into acidophils (staining with acid dye) and basophils (staining with basic dyes). Using immunocytochemical staining techniques, which depend on the ability of antiserum to recognize specific peptides, it has been shown that the basophils are of two types: the somatotrophs, which secrete growth hormone; and the mammotrophs, which produce prolactin. The basophils produce the trophic hormones. The pituitary itself, which weighs about 0.5 g, lies in the bony cavity—the pituitary fossa—and is attached to the hypothalamus by a stalk. Embryologically the infundibulum, which is a downgrowth of the floor of the third ventricle, develops into the pituitary stalk and posterior pituitary. The other ectodermal component is Rathke's pouch, which is a dorsal outgrowth of the buccal cavity. Rathke's pouch eventually detaches itself and develops into the anterior pituitary. TTFFF

MCQ 126

Hypothalamic releasing and inhibiting hormones

1. are peptides or polypeptides.
2. are synthesized only in the hypothalamus.
3. are present in high concentrations in the hypophyseal artery.
4. are in many cases available for clinical use.
5. are highly specific for one pituitary cell type.

Although Geoffrey Harris predicted the presence of hypothalamic hormones many years ago, it was not until the last 15 years that some of the hormones controlling anterior pituitary function have been isolated and synthesized. The first hormone to have its structure identified was thyrotrophin-releasing hormone (TRH), which is a tripeptide. The other releasing hormones are larger, gonadotrophin-releasing hormone (GnRH) or luteinizing hormone releasing hormone (LHRH) being a decapeptide, corticotrophin-releasing hormone (CRH or CRF) having 41 amino acids and growth hormone releasing hormone (GHRH) approximately 40 residues. The exception to the peptide structure seems to be prolactin-inhibiting factor, which appears to be dopamine. A releasing factor has been identified for all anterior pituitary hormones except prolactin, but they are not entirely specific as only one releasing hormone has been described for the gonadotrophins and TRH also stimulates prolactin release. The releasing hormones are synthesized in the parvicellular nuclei in the hypophyseotrophic region of the hypothalamus. The axons terminate in the median eminence, where hormones pass into the pituitary portal circulation and thence to the anterior pituitary. The inferior hypophyseal artery supplies the posterior lobe and the superior hypophyseal arteries supply the median eminence and pituitary stalk. The anterior pituitary receives no direct blood supply, nutrition arriving via the portal vessels. The hypothalamus is not the only site of synthesis of these releasing factors. Many are distributed through the central nervous system and even in other regions, particularly in the gastrointestinal tract. Somatostatin is found in this region and in the pancreas. The effects of the endogenous releasing hormones can be reproduced by the administration of synthetic peptides. This fact can be taken advantage of in testing the anterior pituitary reserve, i.e. the amount of hormone which can be released from the anterior pituitary. Synthetic peptides can also be used therapeutically, as for example in the treatment of fertility. In this case LHRH has to be administered in a pulsatile fashion. Maintained concentrations of LHRH produce decreased gonadal function as a result of down-regulation of pituitary receptors, which blocks gonadotrophin release. TFFTF

MCQ 127

Acromegaly is a disease in which

1. there are low somatomedin concentrations.
2. abnormal growth hormone responses to physiological stimuli occur.
3. symptoms of diabetes mellitus may be present.
4. growth hormone secretion is stimulated by bromocriptine.
5. there is growth of the viscera as well as bony overgrowth.

Oversecretion of growth hormone before the epiphyses have fused results in gigantism, and in acromegaly thereafter. As is implicit in the name the main action of the hormone is to promote growth. It produces extension of the long bone by an action on chondrogenesis and osteogenesis. The action on cartilage is not due to a direct effect of growth hormone, but via the intermediates, somatomedins, a family of peptides of molecular weight 6000–9000. Not only does growth hormone have an effect on bone, but also on tissues, which grow in size to keep pace with the metabolic requirements. Growth hormone has both anabolic and catabolic effects. The anabolic effects are related to growth promotion, enhanced formation of protein and nucleic acids and the retention of other constituents of lean body mass, whereas the effects on fats and carbohydrates are metabolic effects supportive of anabolism and growth. In addition growth hormone has physiological actions which antagonize insulin action. These actions of growth hormone are all reflected in the changes seen in acromegaly. There is progressive enlargement of the hands and feet and growth of certain of the bones of the skull, particularly the lower jaw. There are other features, including arthritis, backache, glucose intolerance and, in 10 per cent of cases, diabetes mellitus. The condition is generally caused by a growth hormone secreting pituitary adenoma. The expanding tumour may produce local pressure effects, including headache and visual field defects. Biochemically the characteristic feature of acromegaly is an elevation of serum growth hormone which is episodic as in normal subjects but does not peak during sleep. Growth hormone is also normally stimulated by stress and suppressed by elevation of blood glucose, but this is not seen in acromegalics. Bromocriptine in normal subjects stimulates growth, but is inhibitory in acromegalics. FTTFT

MCQ 128

Secretion of ACTH in man

1. exhibits a circadian rhythm.
2. occurs in a spurt fashion.
3. occurs in parallel with MSH.
4. is decreased by aldosterone.
5. is stimulated by angiotensin II.

Secretion of the anterior pituitary hormone adrenocorticotrophin (ACTH) is under the control of the hypothalamic releasing hormone CRH or CRF. The release of these peptides is episodic and stimulated by stress such as pain, fever, fear or hypoglycaemia. Probably the most characteristic feature of ACTH secretion is its circadian rhythm. Secretion of CRF and ACTH is controlled by the blood cortisol acting by negative feedback on the CRF-producing neurones and the pituitary corticotrophs. Plasma cortisol is in turn controlled by ACTH, which acts on the adrenal cortex to stimulate steroid synthesis, especially synthesis of cortisol. Administration of ACTH also results in an increase in adrenal blood flow and of corticol protein synthesis, and if administration is prolonged adrenal hypertrophy occurs. ACTH also has effects outside the adrenal gland and when present in excess produces darkening of the skin, as a result of the melanocyte-stimulating hormone (MSH) sequence contained in the first 13 amino acids in ACTH; MSH is not present in the human pituitary. ACTH is synthesized in the form of a large precursor molecule, pro-opiomelanocortin, from which a number of peptides including ACTH and β-endorphin may be cleaved. ACTH itself contains the sequence of α-MSH and corticotrophin-like intermediate lobe peptide (CLIP). Production of excessive amounts of ACTH produces a condition known as Cushing's disease, which is characterized by the effects of excessive steroid production as described in MCQ 134. In this condition ACTH secretion shows an abnormal pattern and does not respond normally to the standard tests. Estimates of ACTH for these tests may be carried out using radioimmunoassay. Bioassays also exist which rely on the action of the hormone on the adrenal gland. Hypophysectomized animals may be used and steroid output from the gland measured. Alternatively dispersed cells or slices of adrenal gland in tissue culture may be used. Lack of ACTH causes failure of secretion from the adrenal cortex, although aldosterone is still produced. TTFFF

In determining anterior pituitary function

1. growth hormone secretion may be diagnosed on the basis of a single plasma estimation.
2. growth hormone deficiency may be established using a glucose tolerance test.
3. hypothalamic LHRH reserve may be determined in the clomiphene stimulation test.
4. ACTH oversecretion may be tested for using dexamethasone.
5. the pituitary thyroid axis may be tested using the insulin tolerance test.

Releasing factors and the anterior pituitary hormones they control are released in a spurt fashion, so that determinations of hormone concentration in a single blood sample are not sufficient to establish over- or undersecretion. Instead challenge tests are used. These are tests to stimulate release if insufficiency is suspected or to inhibit release if oversecretion is believed to be present. The insulin tolerance test is the mainstay of neuroendocrine testing. Its principal use is in the testing of the releasable stores of growth hormone or of the pituitary adrenal axis, although it is also useful in the diagnosis of Cushing's syndrome. The basis of the test is that the neuroglycopenia induced activates the hypothalamic neurones, which in turn, if adequate pituitary reserves are present, stimulate the release of ACTH and growth hormone. Diagnosis of growth hormone excess may be confirmed using a glucose tolerance test, in which growth hormone secretion is suppressed if the levels are elevated as a result of stress, but not in acromegaly. Over-secretion of ACTH characteristic of Cushing's syndrome may be established using the metyrapone test or dexamethasone suppression test. In normal subjects dexamethasone leads to suppression of urinary 11-hydroxycorticosteroids, failure of which is classically seen in Cushing's syndrome. Metyrapone causes a fall in cortisol, leading to an increase in ACTH release and hence in cortisol precursor metabolites in normal subjects and patients with pituitary-dependent Cushing's syndrome. In the latter patients a greatly enhanced cortisol response is seen in the CRF test. The hypothalamic–pituitary–thyroid axis may not be tested with the insulin tolerance test, but rather with the TRH test. In normal subjects with normal circulating T_4 and T_3 levels TRH causes a transient rise in TSH at 20 min with a partial fall to baseline at 60 min. Analysis of an abnormal response will indicate whether the cause of the condition lies in the hypothalamus, pituitary or thyroid.
FFTTF

MCQ 130

In someone with inadequate vasopressin secretion

1. hypertonic sodium chloride infusion will produce a concentrated urine.
2. a dilute urine is excreted.
3. injections of vasopressin can produce cardiovascular effects.
4. plasma osmolality is generally lower than normal.
5. there is general disturbance of the control systems for vasopressin release, including altered function of cardiovascular receptors.

Vasopressin, like oxytocin, is synthesized in the form of a precursor molecule in the hypothalamus and released from the posterior pituitary. Vasopressin has two main sites of action: the kidney and the blood vessels. The hormone acts on the end of the distal convoluted tubule and the collecting duct to increase their permeability to water, which then passes out of the tubule down the concentration gradient into the hypertonic interstitium. Thus an antidiuresis is produced. Vasopressin, as the name suggests, acts on the blood vessels causing vasoconstriction, and in high enough doses may elevate blood pressure. Consistent with the actions of the hormone vasopressin is released in response to an increase in plasma osmolality or to a fall in blood pressure and plasma volume. It is postulated that there are osmoreceptors in the hypothalamus sensitive to changes in osmolality of the extracellular fluid. Hence elevation of plasma osmolality as seen on dehydration or infusion of hypertonic saline leads to vasopressin release and water retention, which returns plasma osmolality towards normal. Similarly a fall in blood pressure, as produced by haemorrhage, stimulates vasopressin release, which in turn causes vasoconstriction and elevates blood pressure. Other actions of vasopressin have been postulated, such as an effect on memory and blood coagulation. Until recently only one condition was known to be associated with disturbance of vasopressin secretion, namely diabetes insipidus, in which vasopressin secretion is generally reduced or absent. There is also a condition known as nephrogenic diabetes insipidus, in which the kidney is unresponsive to vasopressin. In this condition vast quantities of dilute urine are produced so that patients tend to become dehydrated and plasma osmolality elevated. Infusion of hypertonic saline has no effect. Relatively recently the syndrome of inappropriate secretion of vasopressin has been described, characterized by a low plasma osmolality, but relatively concentrated urine. In this condition vasopressin secretion results from causes other than elevated plasma osmolality, so that there is inappropriate water retention. The plasma becomes very diluted, and ultimately water intoxication may result.
FTTFF

Oxytocin

1. is synthesized in the neural lobe.
2. acts together with testosterone and FSH in normal spermatogenesis.
3. stimulates myoepithelial cells surrounding the alveoli and ducts of the lactating breast.
4. action on the uterus is enhanced by oestrogen.
5. in certain conditions stimulates natriuresis.

Oxytocin is a nonapeptide synthesized in the supraoptic and paraventricular nuclei in the form of a precursor molecule. This is packaged within granules and transported to the neural lobe. The neurosecretory granules are stored here until the active hormone can be released into the systemic circulation in response to one of a number of stimuli. The stimuli for oxytocin release are best characterized in the female and are vaginal distension and suckling. Oxytocin does not seem to be important in initiating parturition, but plays an important role in the expulsion of the fetus, acting on the oestrogen-primed uterus to promote contractions. Relatively high oestrogen concentrations are required as oestrogen induces oxytocin receptors in the uterus. Oxytocin is used clinically to aid in labour. Oxytocin is also important in suckling; it causes contraction of the myoepithelial cells surrounding the alveoli and ducts of the lactating breast, thereby causing expulsion of the milk. Oxytocin has a short circulating half-time and therefore has a limited duration of action. Like other pituitary hormones it exhibits 'spurt' release. Oxytocin is found in the male pituitary in similar amounts to those in the female. However, its function in the male is not well established. It may be released during mating and aid in sperm transport. It can also act synergistically with vasopressin to increase sodium excretion by the kidney and hence may contribute to salt and water in the male. There are no known clinical syndromes associated with under-or oversecretion of oxytocin.
FFTTT

MCQ 132

In the thyroid

1. iodide enters by passive diffusion.
2. thiocyanate competitively inhibits the uptake of iodide.
3. hormone synthesis takes place at the cell–colloid interface.
4. TSH stimulates thyroglobulin synthesis and breakdown.
5. TSH stimulates calcitonin secretion.

The thyroid is a bilobed structure, adherent to the front of the trachea and weighing 25 g. It has a very high blood flow, nearly twice that of the kidney, and 50 per cent of the iodine in the blood is removed as the blood passes through the gland. Concentration gradients of 20–100-fold exist, so that much of the daily intake of 1.2 μmol (150 μg) per day enters the thyroid. This active uptake of iodine is produced by an iodide pump linked to an Na^+/K^+ pump. A variety of negatively charged ions with the same van der Waals radii as iodide inhibit the iodide uptake. These include bromide, thiocyanate, pertechnitate and perchlorate. The thyroid gland consists of irregular lobules, each containing 20–40 follicles comprising cuboidal epithelial cells arranged as roughly spherical sacs, the lumina of which contain colloid. This is almost entirely composed of an iodinated glycoprotein called thyroglobulin, from which thyroxine is made. In the synthesis of thyroid hormones the first reaction in the sequence of events after the uptake of iodide is its oxidation. The reaction, which is catalysed enzymically by peroxidase, takes place at or near the surface of the follicular cells. Once formed the iodine immediately attaches itself to the 3-position of the tyrosine ring to form monoiodotyrosine (MIT) and then diiodotyrosine (DIT) is formed. Formation of the iodotyrosines occurs within the thyroglobulin. Organic binding of iodine is followed by coupling of MIT and DIT via an ether linkage to form T_3 and T_4. Before the hormone is released the thyroglobulin–hormone complex is engulfed by cytoplasmic processes and reintroduced into the cell as colloid droplets by endocytosis. Lysosomes then fuse with the colloid droplets and the active hormone is released into the local capillaries. The synthetic processes are stimulated by TSH. FTTTF

Thyroxine and triiodothyronine (T_4 and T_3)

1. circulate in equal concentrations.
2. concentrations increase in pregnancy.
3. concentrations increase on exposure to cold.
4. act by uncoupling oxidative phosphorylation.
5. increase the fluxes of sodium and potassium in responding tissues.

Control of the thyroid gland is via thyroid-stimulating hormone (TSH, thyrotrophin), the secretion of which is controlled by TRH. This is in turn influenced by other areas of the central nervous system, by stress and by thermal changes. The thyroid hormones, which may also be elevated during pregnancy, themselves act at the pituitary to suppress TSH secretion. More T_4 is released from the thyroid than T_3, the total serum concentration of T_3 being 2 per cent of that of T_4. Both hormones are bound to plasma proteins, with only 0.03 per cent T_4 being in the free form as compared to 0.25 per cent T_3. The lower binding of T_3 means that free T_3 is about 30 per cent free T_4. Even though the circulating half-life of T_3 of 1–3 days is shorter than that of T_4, the former is biologically more potent. Thyroid hormones are required for normal growth and development and their actions may be classified as metabolic and developmental. With regard to the metabolic effects, the only definitive action is an increase in basal metabolic rate. This is probably produced by a multiplicity of metabolic effects on the tissues. At one time it was believed that the actions of the thyroid hormones were dependent on uncoupling of oxidative phosphorylation. A more recent suggestion is that the increase in metabolic rate may depend on the *de novo* synthesis of a specific enzyme, possibly Na^+/K^+-ATPase. Thyroid hormones also influence the action of other hormones, clinically the most important action being the effect of thyroxine and catecholamines in increasing heart rate. A lowered blood cholesterol is also characteristic of hyperthyroidism. Hypothyroidism is characterized by opposite effects and if found in the young animal then there is impaired growth and central nervous function. FTTFT

MCQ 134

In Graves' disease

1. plasma TSH concentrations are usually elevated.
2. plasma T_3 concentrations are elevated.
3. immunoglobulins are present, capable of stimulating thyroid tissue.
4. diffuse thyroid hyperplasia is seen.
5. exophthalmos is always seen.

Graves' disease is the most common form of hyperthyroidism. This condition, which results from overactivity of the thyroid gland and increased circulating concentrations of thyroid hormones, especially T_3, is associated with an increase in basal metabolic rate and β-adrenergic activity. Even though appetite may increase there is frequently loss of weight. This is associated with increased heat sensitivity and increased protein catabolism, which may cause muscle wasting. There is enhanced intestinal glucose absorption and a low plasma cholesterol. The patient has an increased heart rate and cardiac output and is restless, anxious and irritable. Eyelid retraction and lag are commonly found. More severely, there can be accumulation of mucopolysaccharides and lymphocytes behind the eye which causes varying degrees of protrusion of the eye, resulting in exophthalmos. Occasionally this can be so severe as to threaten loss of vision as a result of corneal damage or damage to the optic nerve. The elevated activity of the thyroid gland is not due to the elevated TSH concentrations but to the presence of thyroid-stimulating antibodies. These were originally termed long-acting thyroid stimulator because of the slow onset and time-course of activity as compared with TSH. The antibodies bind to the thyroid follicular cell membrane either at or sufficiently close to the thyrotrophin receptor to cause stimulation of adenylate cyclase. This results in hypertrophy of follicular cells without significant destruction of acinar cells. FTTTF

MCQ 135

In the adrenal gland

1. glucocorticoids are synthesized in the zona glomerulosa.
2. testosterone and cortisol are synthesized in both sexes.
3. cholesterol is the precursor of all steroid synthesis.
4. deficiency of 21-hydroxylase leads to congenital adrenal hyperplasia.
5. the metabolic pathway for the synthesis of progesterone and testosterone is different from the gonads.

Steroid hormones are synthesized in the outer cortex, which comprises three zones: a thin outer zone, the zona glomerulosa, where aldosterone is produced; a wide middle zone, the zona fasciculata; and the inner zona reticularis. The glucocorticoids cortisol and corticosterone are synthesized in these last two

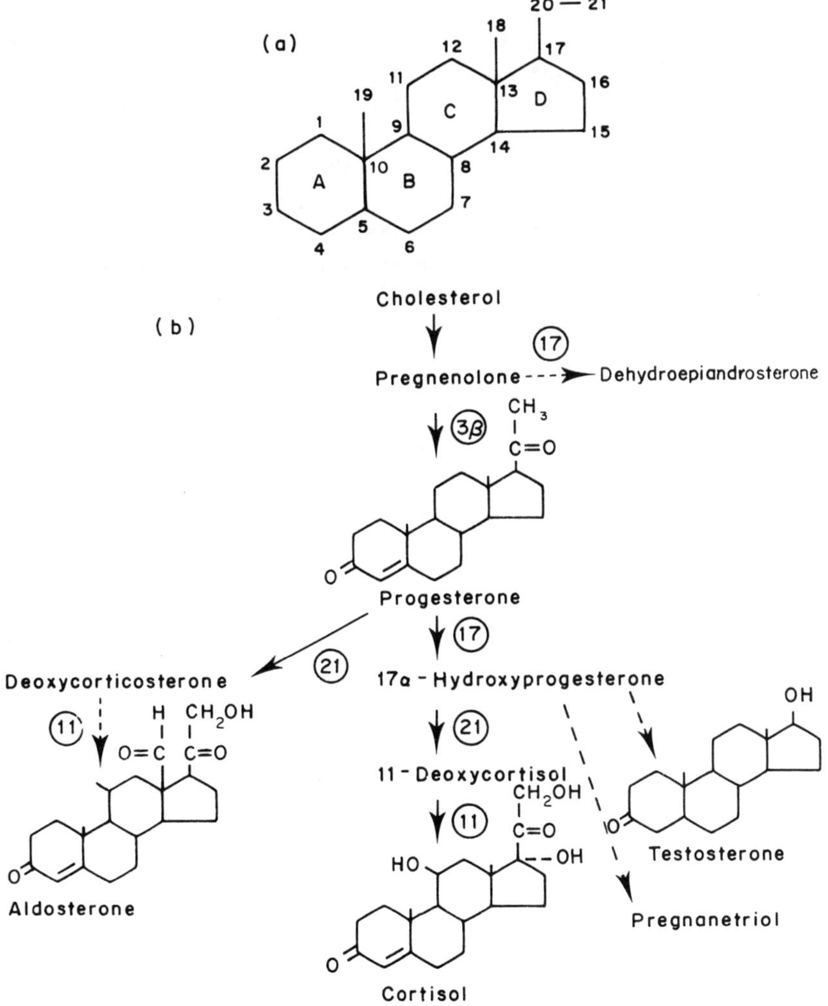

Figure (a) Basic structure of steroid molecule showing labelling of ring and numbering of carbon atoms. (b) An outline of steroid synthesis showing (in circles) the enzymes involved: 17 represents 17α-hydroxylase; 3β, 3β-dehydrogenase; 21, 21-hydroxylase; 11, 11β-hydroxylase. A dotted arrow indicates that an intermediate has been omitted.

zones. Steroid hormones are derivatives of sterane comprising four rings designated A, B, C and D. Each carbon atom on the rings is also given a number. The sterane can be extended by the addition of methyl groups on C(10) and C(13) and by a side chain on C(17). Steroids with 18 carbon atoms are oestrogens, those with 19 are androgens and those with 21 are corticosteroids or progestagens. In the synthesis of steroids pregnenolone is produced in the mitochondria from cholesterol by side-chain cleavage at position C(20/22) under the action of cholesterol desmolase. The pregnenolone is transported back to the smooth endoplasmic reticulum, where it is converted to progesterone by Δ^5-3-hydroxysteroid dehydrogenase and then hydroxylated at either C(17) or C(21). The latter derivative (11-deoxycorticosterone) is hydroxylated in the mitochondria to corticosterone. The 17α-hydroxylation of progesterone leads to the production of cortisol. In the zona glomerulosa it is converted to aldosterone by successive hydroxylation and dehydrogenation at the C(18) position. In the zona reticularis pregnenolone is converted in the smooth endoplasmic reticulum to 17β-hydroxypregnenolone and the side chain at C(17) cleaved to produce a C(19) steroid, dehydroepiandrosterone, which is dehydrogenated to androstenedione. If there is an enzyme deficiency, the precursor tends to accumulate and the effects of the precursors will be manifest. There are also low circulating concentrations of cortisol, so that ACTH release is enhanced, which causes adrenal hyperplasia. The commonest deficiency is of 21-hydroxylase, which leads to abnormally high concentrations of 17-hydroxyprogesterone and testosterone, so that females are virilized and males develop pseudoprecocious puberty. FTTTF

MCQ 136

Aldosterone

1. production is stimulated by a fall in renal blood flow.
2. secretion may be influenced by potassium ion concentration.
3. causes sodium loss and potassium retention in sweat and saliva.
4. when administered continuously to normal subjects for over two weeks causes an escape mechanism to operate.
5. when excreted in excess leads to muscle weakness and hypertension.

Aldosterone is a C(21), 18-hydroxysteroid produced by the zona glomerulosa in amounts of about 400 nmol day^{-1}. It is the most potent mineralocorticoid produced by the adrenal cortex. The main site of action of aldosterone is the kidney, in the distal tubule and collecting duct, where it promotes reabsorption and hence retention of sodium. This in turn is linked to increased excretion of potassium and hydrogen ions. The sodium retention leads to increased plasma volume and an increase in blood pressure. There may also be an effect on sodium flux in arteriolar muscle. Aldosterone, being a steroid, acts at a nuclear level to stimulate synthesis of a protein, in this case one which promotes sodium transport. There is thus a delay of about one hour before the effect of aldosterone is seen. If aldosterone administration is continued, then after one to three weeks excess sodium is once again excreted. This 'escape' phenomenon probably results from readaptation of the mechanism responsible for the reabsorption of filtered sodium in the proximal tubule. Aldosterone secretion is largely controlled by angiotensin II. Potassium and ACTH in very high concentrations also have an effect. Angiotensin II is formed by the action of converting enzyme in the lungs, the substrate being angiotensin I. This is produced from angiotensinogen in the plasma through the action of renin. This proteolytic enzyme is produced in the juxtaglomerular apparatus of the kidney, as described in MCQ 91. Three main factors control the secretion of renin: changes in the activity of the sympathetic nerves, the flux of sodium across the macula densa of the distal tubule and finally changes in transmural pressure. Thus when sodium flux is high, as when sodium is plentiful, or there is an increase in pressure, renin secretion is suppressed, so that aldosterone secretion is reduced and sodium is lost. Renin production is inhibited by a fall in renal plasma flow.
TTFTT

MCQ 137

In Cushing's disease increased glucocorticoid activity leads to

1. increased protein catabolism resulting in muscle wasting.
2. increased redistribution of fat resulting in central obesity.
3. involution of thymic tissue with susceptibility to infection.
4. increased secretion of acid in the stomach and predisposition to gastric ulcer.
5. suppression of ACTH.

Cushing's disease is due to excess secretion of ACTH. Many patients have basophilic microadenomas and a few have larger tumours. The abnormality may also lie at the level of the hypothalamus, resulting in excessive CRF production. The major clinical signs are obesity with a 'buffalo' hump, purple striae, excessive bruising, hypertension, hirsutism, psychiatric disturbances, polyuria, oligomenorrhoea (irregular menstruation), muscle weakness, osteoporosis and spontaneous fractures. These changes reflect the increased mineralocorticoid activity. Cortisol has pronounced effects on carbohydrate metabolism, leading to increased gluconeogenesis, and hepatic glycogenolysis, leading to diabetes mellitus. It also has permissive effects, which include the effect of noradrenaline on carbohydrate metabolism. The muscle wasting and easy bruising, thin skin and osteoporotic bones result from the increased protein catabolism. The effect on fat metabolism leads to an increase in redistribution of body fat, with central obesity, moon face, 'buffalo' hump and relatively thin limbs. In addition to its effects on metabolism cortisol acts on the body's defence mechanisms, suppressing the tissue response to injury and having an anti-inflammatory effect. It can also produce involution of lymphatic and thymic tissue. All these effects lead to susceptibility to infection. Amongst the other actions are a mineralocorticoid effect. Normally this is not significant, but if large amounts of cortisol are produced then electrolyte disturbances may be seen, with a fall in plasma potassium. Still further actions are an effect on the central nervous system to produce euphoria and stimulation of acid secretion by the stomach, resulting in a predisposition to gastric ulcers. In Cushing's syndrome, in which the primary cause is an adenoma or carcinoma of the adrenal cortex, the high concentrations of glucocorticoids will suppress ACTH release. TTTTF

MCQ 138

Adrenaline, in contrast to noradrenaline

1. is the predominant secretion of the adrenal medulla.
2. is predominantly a vasoconstrictor.
3. when infused intravenously promotes bradycardia.
4. increases blood glucose concentrations to a greater extent.
5. is a more effective dilator of the bronchi.

The catecholamines adrenaline and noradrenaline are synthesized in the adrenal medulla. The adrenal glands are paired glands, triangular in shape and positioned on the superior poles of the kidney. The inner medullary core of modified ganglion cells is derived embryonically from the neural crest and the cells are termed phaeochromocytes, while the surrounding cortex comprises epithelioid cells. Adrenaline is formed from noradrenaline through the action of phenyl-ethanolamine-*N*-methyltransferase, an enzyme which requires the presence of a high concentration of corticosterone. Noradrenaline is in turn synthesized from tyrosine in three steps via DOPA and dopamine. The first step in the conversion of tyrosine to DOPA is the rate-limiting step. About 80 per cent of the resting catecholamine secretion is adrenaline. Catecholamine release is under the control of and influenced by many stimuli, including emotion, hypoglycaemia and trauma. Catecholamines have a very short circulating half-time in the plasma, being metabolized and excreted in the urine or taken up in the nervous tissue. Adrenaline causes vasoconstriction in the skin and gastro-intestinal tract, and vasodilatation in the vessels to the skeletal muscle, so that a drop in blood pressure would result were it not for an increase in the heart rate and the force of contraction. Noradrenaline too produces an increase in blood pressure. Adrenaline also reduces gut motility, causes bronchodilation and piloerection. Noradrenaline has a much less pronounced effect in reducing gut motility and does not cause bronchodilation. Adrenaline also has metabolic effects, promoting hepatic glycogenolysis leading to hyperglycaemia, and increasing the plasma concentrations of free fatty acids. Noradrenaline is much less potent in influencing carbohydrate metabolism, but has a more potent effect on fat metabolism. The difference between the actions of the two hormones is explained in part by the presence of different receptor populations: α and β receptors. The β adrenoreceptors can be divided into β_1 and β_2 receptors. Noradrenaline reacts only with α and $\beta_2$receptors and so cannot cause broncho-dilation and is less potent on the gut. Likewise adrenaline can produce vaso-constriction in vessels with α and β_2 receptors. The α receptors also fall into the sub-groups α_1 and α_2. TFFTT

MCQ 139

Calcium

1. is present in the body in amounts of 500 g (12.5 mol).
2. is important in blood-clotting.
3. exists only in the free form in plasma.
4. increases the excitability of the neuromuscular junction.
5. daily intake is 25 mmol.

There is a total of 1 kg (25 mol) of calcium in the body, of which 99 per cent is present in bone, 5 g being present in exchangeable bone. Approximately 1 g (25 mmol) is found in the extracellular fluid and 10 g (250 mmol) in the intracellular compartment. There is a continuous exchange between the different calcium pools, but balance between them is generally maintained. Of the 25 mmol calcium taken in daily about 20 mmol are lost in the faeces. Some 12.5 mmol (500 mg) are daily secreted into the intestine and 15 mmol are reabsorbed. Of the 0.25 mol filtered daily by the kidney, 2.5 mmol are excreted. About 10.0 g daily exchange with the exchangeable pool of calcium in the bone. Bone consists of 65 per cent inorganic crystals, primarily hydroxyapatite, $3Ca_3(PO_4)_2.Ca(OH)_2$ and 35 per cent organic materials. Bone has a number of functions, the most important being to support the body and protect vital organs, including the brain and bone marrow. It acts as a reserve of calcium and phosphorus and contributes to their balance in the body. Plasma calcium exists in three forms. Just under half is in the ionized form, which is the metabolically active form, about 46 per cent is bound to protein, largely albumin, and the remaining 6 per cent is complexed to small organic molecules, e.g. citrate. Calcium in the plasma and extracellular fluid is kept constant, being regulated by alterations in bone turnover, changes in absorption from the diet and by alteration in the renal excretion. These mechanisms are controlled by three factors, namely parathyroid hormone, calcitonin and vitamin D. Calcium is necessary for the metabolism of all cells and for the formation of teeth as well as bones. It is important in the activity of excitable tissues, acting as a membrane stabilizer and being involved in action–contraction coupling in muscle contraction, in stimulus–secretion coupling at nerve terminals and in the functions of cardiac muscle. It is needed for the stimulus–secretion coupling in endocrine glands, the activation of some enzymes and blood coagulation. Phosphate metabolism is closely linked to that of calcium. FTFTT

MCQ 140

With regard to the parathyroid gland and 'C' cell (calcitonin cell) system

1. the parathyroid gland is of endodermal origin.
2. the 'C' cell system is of neural crest origin.
3. calcitonin-producing cells in man are only found in the thyroid gland.
4. it produces polypeptide hormones.
5. perfusion of the 'C' cell system with a solution of high calcium concentration results in discharge of the neurosecretory granules.

With regard to the calcium-regulating hormones, calcitonin is produced in the 'C' cells of the thyroid gland and parathyroid hormone in the four parathyroid glands on the posterior surface of the thyroid gland. In man calcitonin is also present in the parathyroids and thymus as well as in the thyroid, so that no significant changes in bone metabolism are seen on complete thyroidectomy. The parafollicular or 'C' cells of the thyroid are distinct from the follicular cells, being derived from the neural crest from which they migrate to the last bronchial pouch. The 'C' cells possess a number of cytochemical characteristics which are used as a paradigm for a class known as the APUD system—amine precursor uptake decarboxylase. The system includes the adrenal medulla and other peptide-secreting cells. The superior parathyroids arise from the dorsal portion of the fourth brachial pouch and migrate caudally together with the thyroid to reach the final position. The inferior parathyroids arise with the thymus from the third brachial pouch. These complete embryonic origins may result in biochemical abnormalities. Both parathyroid hormone and calcitonin are peptide hormones. Parathyroid hormone is a single chain of 84 amino acids, while calcitonin is a single chain of 32 amino acids. The major factor in the secretion of both hormones is the plasma calcium concentration. Hypercalcaemia suppresses secretion of parathyroid hormone, while hypocalcaemia, which can be produced by the administration of agents such as EDTA which complexes calcium ions, stimulates secretion. In contrast, hypercalcaemia stimulates calcitonin. Infusion of calcium into an experimental animal induces 'C' cell degranulation. Other factors which may modulate parathyroid hormone secretion but whose physiological significance is not clear are adrenaline and derivatives of vitamin D. TTFTT

MCQ 141

The concentration of ionized calcium in plasma will rise

1. if calcitonin is given.
2. if increased calcium is ingested.
3. if the concentration of inorganic phosphate is raised.
4. if large doses of vitamin D are given.
5. with voluntary hyperventilation.

The normal level of plasma calcium is 2.3–2.6 mmol/l. As described in MCQ 138 plasma calcium is partly bound and partly free. The protein binding depends not only on the plasma protein concentration but also on the plasma pH: the higher the pH, the greater the amount of protein anion available to bind calcium. Plasma calcium ion concentration is regulated by the interplay of three hormones: parathyroid hormone, calcitonin and vitamin D. Parathyroid hormone and vitamin D elevate plasma calcium ion concentrations, while calcitonin lowers them. There are three main sites at which they could act to influence plasma calcium: absorption from the gastrointestinal tract, formation and resorption of the bone and excretion from the kidney. The main actions of parathyroid hormone are on the bone and the kidney. In bone its effect is to increase resorption, probably through an effect on the osteoclasts. The effect on the kidney is a composite one. The hormone stimulates reabsorption of calcium in the tubules, so that urinary excretion is reduced, but there may also be a rise in renal excretion as a result of the greater filtered load. The hypocalcaemic effect of calcitonin depends primarily on its ability to inhibit the mobilization of calcium from bone by suppressing the activity of the osteoclasts. It may also have a stimulatory effect on bone formation. The main action of vitamin D, which is produced by 1,25-dihydroxycholecalciferol, is to stimulate calcium absorption in the small intestine, where it acts on the intestinal mucosa like a steroid hormone. It is believed to stimulate formation of a calcium-binding protein important in transporting calcium across the intestinal cell. FFFTF

MCQ 142

In the pancreatic islets are

1. α cells which produce insulin.
2. β cells which produce glucagon.
3. δ cells which secrete a 14 amino acid peptide.
4. β cells whose secretion is reduced by endogenous catecholamines.
5. α cells whose secretion is reduced by D-glucose.

The pancreas arises from the dorsal and ventral endodermal outgrowths from the foregut. The majority of the gland comprises exocrine cells which secrete enzymes into the pancreatic duct. Scattered amongst these cells are aggregates of endocrine cells, which represent about 2 per cent of the pancreas. Three major cell types are found in the islets: α, β and δ. The α cells synthesize proglucagon, from which is split off glucagon, a single-chain peptide comprising 29 amino acids. Pre-proinsulin is synthesized in the β cells. From this large peptide a 24 amino acid chain is split off to give proinsulin, a single-chain peptide, which depending on the species comprises 81 or 86 amino acids. This is then packaged into granules where it is converted into insulin by cleavage of the C-terminal peptide. Insulin itself has a molecular weight of 6000, with 21 amino acids in the A chain and 30 in the B chain, the chains being linked by two disulphide bridges. The δ cells contain somatostatin in their granules. The main regulator of insulin and glucagon secretion is the circulating concentration of glucose. A rise in blood glucose stimulates release of insulin but suppresses that of glucagon. Amino acids stimulate the release of both hormones, while somatostatin inhibits release of insulin. There is also neural control of pancreatic hormone release. The stimulation of the vagus or pancreatic nerve promotes release of insulin, a response which can be blocked by atropine. There are also α- and β-adrenergic receptors. Sympathetic adrenergic nerve stimulation releases glucagon via α receptors. Insulin release is stimulated via α receptor stimulation (noradrenaline) and inhibited by β receptor stimulation (adrenaline). Several of the gastrointestinal hormones, such as secretin, gastrin, CCK/PZ and GIP (glucose-dependent insulinotrophic hormone), stimulate insulin release. FFTTT

MCQ 143

Hypoglycaemia

1. may be prevented by gluconeogenesis.
2. may be caused by increased secretion of insulin.
3. is accompanied by an increased secretion of catecholamines, growth hormone and cortisol.
4. occurs after 48-h fasting.
5. may be caused by hypopituitarism.

Hypoglycaemia may occur in many circumstances, but glucose levels rarely fall below 2.5 mmol l^{-1}, when symptoms are seen. Nearly all hypoglycaemic states are intermittent. The symptoms represent a combination of the effects of glucose lack and adrenaline release. Hypoglycaemia produces irritability, agitation, confusion and sleepiness, while adrenaline produces nervousness, weakness, palpitations, hunger, fear and sweating. The causes of hypoglycaemia are numerous. A common cause is over-production of insulin by tumours or over-secretion of the hormone. Insulin is the only hormone which reduces circulating glucose concentrations and it does so in a number of ways. In contrast to most hormones concerned with energy regulation it is purely an anabolic hormone which tends to spare fat and protein so long as there is sufficient carbohydrate to meet metabolic needs. It facilitates uptake of glucose into tissues, muscle and adipose tissue being the most important. It also stimulates the production of glycogen from glucose, inhibits the breakdown of liver glycogen and depresses gluconeogenesis. It promotes transport of amino acids into liver and muscle cells, a process which differs from the transport of glucose in that amino acid transport is an active process linked to the Na^+/K^+ pump. The overall action of insulin is to promote synthesis and storage of material so that, for example, fatty acid synthesis from acetyl coenzyme A is increased at the expense of ketone body production. Maintenance of blood glucose concentrations is important and stable levels are found even after 48-h fasting. While there is one hormone which lowers blood glucose there are a number which raise it, including catecholamines, growth hormone, cortisol, glucagon and thyroid hormones. These hormones are released in response to hypoglycaemia. TTTFF

MCQ 144

In diabetic ketoacidosis

1. serum osmolality may exceed 330 mOsm kg^{-1}.
2. the fluid deprivation is 5–6 l.
3. the most common occurrence is in patients with juvenile onset diabetes mellitus.
4. the pH of the blood is usually below 7.0.
5. hyperventilation is seen.

There is no agreed definition of diabetes mellitus. It has classically been considered to be a disease of glucose over-production by the liver and under-utilization by insulin-dependent tissues. It occurs in two main forms: early onset (under 30 years of age) and late onset. The former is characterized by weight loss, thirst, polyuria, nausea and vomiting, and insulin is needed in treatment. The polyuria results from the presence of glucose in the urine, which spills over from the plasma as the concentrations are so great. The polyuria in turn leads to marked fluid deprivation and polydipsia. The high glucose concentrations and fluid loss can result in plasma osmolalities of over 300 mOsm kg^{-1}. Patients with late-onset diabetes are usually obese and complain of thirst and polyuria. However, the illness is usually mild and is unresponsive to insulin. Ketoacidosis may be a prominent feature of untreated early-onset diabetes, but may also occur when the diabetes escapes from control, as with stress resulting from infection or surgery. In diabetes fatty acids as well as ketone bodies are used as an energy source in lieu of glucose. The fatty acids released on rapid mobilization of triglycerides are taken to the liver and metabolized. As a result, the ketone bodies, β-hydroxybutyric acid and acetoacetic acid are produced in excess of the ability of the tissues to use them, so that their circulating concentration rises. This, in addition to producing nausea and vomiting, results in a serious disturbance of acid–base balance. The acidosis stimulates the respiratory centre, producing hyperventilation; the carbon dioxide is blown off and respiration depressed until carbon dioxide accumulates, so that sighing respiration is produced—Kussmaul breathing. TTTFT

MCQ 145

Spermatozoa

1. contain 23 chromosomes.
2. are stored in the seminal vesicles.
3. mature more rapidly at 37 °C than at 35 °C.
4. need FSH for normal development.
5. are required in a concentration of 10^6 ml^{-1} for fertilization.

Spermatozoa are produced in the seminiferous tubules of the paired testes. The seminiferous tubules possess two types of cells: the spermatogonia or germ cells from which the spermatozoa are derived; and Sertoli cells, which are necessary for spermatogenesis and are sometimes called germ cells. Spermatogenesis within the germinal epithelium takes about 10 weeks. Spermatogonia divide repeatedly by mitotic division to give reserve spermatogonia and a second type which become primary spermatocytes. These then undergo the first meiotic division, forming secondary spermatocytes which contain only 23 chromosomes. A second meiotic division produces the spermatids, which gradually transform into spermatozoa. At this stage the spermatozoa are not capable of independent motility or fertilization and undergo further maturation during their passage through the epididymis. From the epididymis the vas deferens leads to the urethra. A sperm count of less than 20×10^6 ml^{-1} leads to infertility. Sterility is also produced by exposure to high temperatures. In many mammals, including man, the testes are carried in a scrotal sac, which maintains the temperature a few degrees lower than the core temperature. Failure of the testes to descend leads to degeneration of the spermatogenic epithelium. For the initiation and maintenance of spermatogenesis, the presence of the two gonadotrophins is required. Follicle-stimulating hormone is necessary for the initiation of spermatogenesis, while LH stimulates synthesis of testosterone by the Leydig cells. Testosterone maintains spermatogenesis but cannot initiate it. TFFTF

MCQ 146

Testosterone

1. is produced by the Sertoli cells.
2. production is controlled by LH.
3. is absent in the embryo.
4. can act indirectly on the prostate through conversion to dihydrotestosterone.
5. produces a negative nitrogen balance.

The Leydig cells of the testes produce a number of androgenic steroids, the most important of which is testosterone. The others are androstenedione and dehydroepiandrosterone. Synthesis of these hormones proceeds via cholesterol and pregnenolone in the same way as adrenal androgens. Testosterone production is controlled by LH. Testosterone is produced in the fetus. The human hypothalamus synthesizes LHRH by the 8th week of gestation, whereas the gonadotrophins are produced by the 10–13th week. Testosterone production follows. In the fetus the androgens are responsible for the differentiation of the male genitalia. During puberty they are responsible for the growth and epiphyseal fusion, production of male skeletal proportions, enlargement of the larynx and breaking of the voice. They produce the male hair pattern, enlargement of the penis, scrotum with folds and prostate, as well as giving increased mascularity and strength. In the adult they serve to maintain masculinity as well as being responsible for the development of libido and potency. Testosterone has a widespread effect and elevates the metabolic rate. It promotes the retention of nitrogen and electrolytes and stimulates protein synthesis. In androgen-dependent tissues such as the prostate and seminal vesicles, testosterone exerts its effect via the 5α reduction product, dihydrotestosterone, which is formed in the responsive cells themselves. The majority of testosterone in the plasma is bound to globulin — sex binding globulin — which also carries oestrogen. As with adrenal medullary steroids, testosterone is largely degraded in the liver and conjugated with sulphate or glucuronic acid before excretion in the urine as 17-oxosteroid. TTFTF

MCQ 147

In a normal healthy woman of 25 years the ovaries

1. each contain over 100 000 oocytes, which are present at birth.
2. produce oestrogenic hormones chiefly in the granulosa cells.
3. each cycle have a number of ova which start to develop.
4. require high concentrations of LH during the luteal phase.
5. contain ova with 23 chromosomes.

The adult ovaries weigh about 7 g each and are attached by ovarian ligaments to the back of the broad ligaments. Microscopically the ovary has a capsule of connective tissue—the tunica albuginea—and within a cortex containing follicles embedded in supportive tissue called the stroma. Formation of ova commences in the fetal ovary. At the 11–12th week of gestation the oogonia begin the first meiotic division, but this is arrested without the division being completed. At this stage they are called primary oocytes and form part of the primordial follicles. The number of primordial follicles reaches a peak of 7×10^6 between 20 and 28 weeks gestation. By birth the number has fallen to 2×10^6 and by puberty it is only 300 000. Primary oocytes remain in a state of arrested development from the time at which they are formed until just before ovulation. Just before ovulation the mature ovum divides. The formation of the ovum is similar to that of the primary spermatocyte in that during a meiotic division each daughter cell receives 23 chromosomes. However, one cell retains virtually all the cytoplasm. At the beginning of each menstrual cycle a cohort of 20 early follicles enlarge with proliferation of the flattened follicle cells to form secondary follicles, but only one follicle matures fully. The ovum matures into a Graafian follicle under the influence of gonadotrophins. The follicle grows until it ruptures, extruding the ovum, and the other follicles degenerate. At the time of ovulation the follicle fills with blood and is then converted to the corpus luteum, where oestrogen and progesterone are both synthesized. Ovulation is brought about by a surge in the secretion of LH and to a lesser extent FSH. The rise in plasma gonadotrophin concentrations is believed to lead to increased synthesis of plasminogen activator in the follicle which aids follicle rupture and release of the ovum. TFTFF

Over the course of the menstrual cycle

1. oestrogen dominates the proliferative phase of the endometrium.
2. the loss of menstrual fluid is about 30 ml.
3. cervical mucus becomes more viscous at the time of ovulation.
4. menstruation is preceded by an increase in circulating concentrations of ovarian steroids.
5. menstruation is associated with dilatation of the basal segment of the spiral arteries within the endometrium.

The changing ovarian steroid concentrations over the menstrual cycle produce changes in the female genital tract. Over the first half of the cycle, termed proliferative (regarding uterine changes) or follicular (concerning ovarian changes), there is an increase in the plasma oestrogen concentration. This steroid is produced in the ovum under the influence of LH and then transported to the granulosa cells, where it is aromatized to oestrone and oestradiol. The oestradiol stimulates the proliferation of the endometrium and synthesis of receptors for oestradiol and progesterone in its cells. It also stimulates secretion of clear mucus from the cervix, which allows sperm penetration, together with the maturation of the vaginal epithelium. In addition it feeds back on the hypothalamus and pituitary. Normally it exerts negative feedback but if sufficiently elevated concentrations are maintained for 38 h then positive feedback occurs and the LH surge follows. After ovulation during the second half of the cycle — called in terms of uterine changes the secretory phase, or in terms of the ovarian changes the luteal phase — both oestrogen and progesterone are produced by the ovary. Progesterone acts on oestrogen-primed tissues to produce a number of changes, including the production of a secretory endometrium and the secretion of thick cervical mucus with leucocytes. The changes thus render the endometrium suitable for implantation of a fertilized ovum. Progesterone also causes a rise in body temperature, a fact which may be used in determining the time of ovulation. If fertilization and blastocyst implantation do not occur then the corpus luteum begins to involute and steroid formation fails. The maintenance of the secretory epithelium is then no longer possible, the spiral arteries collapse and the endometrium sloughs off and menstrual bleeding occurs, the loss of menstrual fluid being about 30 ml. TTFFF

MCQ 149

Oestrogenic hormones

1. are chiefly bound to albumin in the plasma.
2. decrease the response of the uterus to oxytocin.
3. produce an increase in body temperature.
4. produce hypertrophy of the myometrium.
5. help to produce mammary duct proliferation at puberty.

The principal oestrogenic hormone in the human is oestradiol, but oestrogen is present in similar concentrations. Oestradiol in the plasma is largely bound to sex hormone-binding globulin, with some binding to albumin. Oestrogenic hormones have changing roles throughout the reproductive life-span. At puberty they are responsible for the growth spurt, epiphyseal fusion and broadening of the pelvis. They also stimulate growth of the uterus and breasts and determine the female figure by controlling the deposition of fat. In addition they are responsible for the development of the external genitalia and growth of the pubic and axillary hair. Together with progesterone they initiate and maintain menstruation. During pregnancy oestrogens cause growth of the breast duct system and myometrial hypertrophy, together with fluid retention and an increase in uterine blood flow. In the ovum oestrogen synthesis commences in the theca cells of the follicle and is completed in the granulosa cells, where the precursors are aromatized to yield oestrone and oestradiol. During the first three months of pregnancy oestradiol is synthesized in the corpus luteum. Thereafter it is produced in the fetoplacental unit. Around the ninth week of pregnancy the fetoplacental unit develops cholesterol desmolase activity so that it can synthesize progesterone and pregnenolone, which then crosses into the fetus. Here it is converted to dehydroepiandrosterone, which crosses back and is transformed to oestriol in the placenta. Plasma oestradiol concentrations rise rapidly before parturition, which causes the induction of oxytocin receptors, hence rendering the uterus sensitive to oxytocin. FFFTT

MCQ 150

In women during the first three months of pregnancy

1. progesterone is produced by the fetoplacental unit.
2. chorionic gonadotrophin maintains the corpus luteum.
3. increasing circulating concentrations of prolactin are observed.
4. the uterine muscle is kept quiescent by high concentrations of progesterone.
5. human placental lactogen is present in higher concentrations in plasma than during the last three months.

Conception requires that the spermatozoa should fertilize an ovum in the Fallopian tube. The fertilized ovum is transported down the Fallopian tube, during which time the blastocyst stage, comprising some 16 cells, is reached. After successful implantation of the blastocyst in the endometrium, it continues to develop and the trophoblast begins to secrete human chorionic gonadotrophin (HCG), a glycoprotein with LH-like activity. This hormone is secreted into the maternal plasma and its detection in plasma, or more usually in urine, is used as the basis of early pregnancy testing. Early-morning urine is used as it is most concentrated. The HCG is usually detected using agglutination of inert particles such as latex, induced by the antibody–antigen reaction. Concentrations of HCG increase up to about 20 weeks gestation. In the early stages of pregnancy production of HCG sustains oestradiol production by the corpus luteum and thus supports the fetoplacental unit. The placenta gradually takes over after 8–9 weeks and then becomes the principal site of oestrogen and progesterone synthesis. Concentrations of progesterone rise steeply at the beginning of pregnancy, fall away for 40–60 days and then increase up to the time of delivery. The high progesterone concentrations help to keep the uterus quiescent. One of the major hormones synthesized by the placenta is human placental lactogen (HPL) or human chorionic somatomammotrophin, a simple protein composed of a single chain of about 190 amino acids cross-linked by two disulphide bridges. Plasma HPL concentrations increase throughout pregnancy until the last few weeks, when they flatten off. The related pituitary hormone prolactin also increases in pregnancy. Other hormones synthesized by the placenta include two thyroid-stimulating agents. FTTTF

MCQ 151

Towards full term of pregnancy in the human

1. the blood volume is higher than at 20 weeks of pregnancy.
2. blood flow through the brain is higher than at 6 weeks of pregnancy.
3. the supine position can lead to hypotension.
4. glomerular filtration rate is increased as compared with the initial weeks.
5. alveolar ventilation is higher than at 6 weeks of pregnancy.

During pregnancy there is an increase in weight of about 12.5 kg, the fetus, placenta, etc. representing 5 kg of this. Among the changes in the mother during pregnancy is a progressive increase in the blood volume from approximately 4.0 l to 5.4 l. This is said to be an adaptation to meet the demands of the uterus and fetoplacental unit and to provide enhanced flow to the skin for the elimination of heat and to the kidneys for the excretion of additional waste products. Both plasma volume and packed cell volume increase, but the increase of the former is greater. This increase in blood volume is accompanied by an increase in cardiac output. There is, however, a small fall in systolic pressure and a greater fall in diastolic pressure. The developing contents of the uterus may exert pressure on the veins returning from the legs, thus reducing venous return. This can be so marked in the supine position as to lead to marked hypotension. In parallel with the increase in cardiac output there is increased blood flow and glomerular filtration rate. There is increased sodium and water retention, which has been suggested to result from the action of the steroidal sex hormones. Of the 3.5 l retained in the body, the increased plasma volume represents some 1.3 l. The upward displacement of the diaphragm might be expected to reduce vital capacity, but there is an increase in the width of the cavity so that the vital capacity is unchanged. There is, however, an increase in the pulmonary ventilation which may be due to the greater sensitivity of the respiratory centre to carbon dioxide. It is especially in the middle trimester of pregnancy that there is hypochlorhydria and decreased motility in the stomach. In the early months of pregnancy there is nausea and vomiting in the morning, but the cause is not clear. It could be due to the effect of steroids on gastrointestinal smooth muscle. Changes in steroid hormone concentrations may also be responsible for the changes in mood seen. TFTTF

MCQ 152

In the fetus

1. the rate of growth is about 160 g per week in the last 5 weeks of pregnancy.
2. glucose is the major energy substrate.
3. free fatty acids are readily transported to the circulation from the maternal blood across placental cells.
4. fetal haemoglobin has a lower affinity for oxygen than that of the adult.
5. pressure in the pulmonary artery is greater than in the aorta.

In addition to being an endocrine gland, as described in MCQ 149, the placenta serves to transfer substances between the mother and fetus. Thus the placenta serves as lung, kidney and gastrointestinal tract for the fetus. It needs to function efficiently as, although the initial rate of placental growth is greater than that, it later slows down. The fetus weighs approximately 3.5 kg at full term, the rate of growth being about 160 g per week. Glucose is the major energy substrate. Amino acids and fatty acids, in addition to some carbohydrates, are readily transferred across the placenta. The PO_2 of blood supplying the fetus is low when compared with that of the arterial blood of the adult, but a number of factors improve the efficiency with which oxygen is transported to the tissues. First, fetal haemoglobin, as described in MCQ 74, has a higher affinity for oxygen than adult haemoglobin; second, there is a greater haemoglobin concentration in the fetus, and third, the uptake of oxygen by the fetal blood is aided by the double Bohr effect. The oxygen dissociation curve is affected by the pH, being shifted to the left with a fall in pH. This effect is greater in the fetus. The pH in the placenta is low, but the oxygen content of the fetal blood at this pH is greater than the maternal. As the placenta serves a number of functions, certain tissues in the fetus may be bypassed. Most of the blood returning to the fetus via the umbilical vein bypasses the liver in the ductus venosus and travels to the right atrium. Blood does not pass through the lung, but instead reaches the left side of the heart via the foramen between the right and left atrium and ductus arteriosus which transfers blood directly from the pulmonary artery to the aorta. The blood does not pass through the lungs, as the resistance of the collapsed lungs is high, the pulmonary arterial pressure being 5 mm greater than that in the aorta. Blood travels to the placenta via the umbilical artery and so bypasses the digestive tract. TTTFT

MCQ 153

For lactation to occur in women

1. growth hormone and cortisol together are required to produce duct growth.
2. progesterone concentrations must fall.
3. LH secretion is essential.
4. increased prolactin secretion is required.
5. oxytocin is required for milk ejection.

At birth the female breast consists of atrophic ducts and no alveoli. At puberty there is growth of the ducts and connecting fibrous tissue, increased deposition of fat and development of the alveoli. Oestrogen stimulates the development of the duct system by causing elongation and thickening of the lactiferous ducts. Progesterone stimulates alveolar growth and leads to the appearance of buds at the distal ends of the ducts, which differentiate into alveoli. Pituitary mammotrophic factors are also necessary for normal breast development. They include prolactin and possibly growth hormone and the pituitary gonadotrophins. Prolactin is the major lactogenic factor in the human. When the breast is fully developed under the influence of the ovarian steroids, prolactin results in further alveolar formation and the secretion of milk. During pregnancy the breasts enlarge and develop further, mainly in response to the placental hormones present in the circulation, namely oestrogen, progesterone and placental lactogen. Lactation is, however, inhibited by the high concentrations of oestrogen and progesterone. Post partum, the onset and maintenance of lactation is immediately due to prolactin. The breast at first secretes colostrum after delivery. It has a higher protein content than milk, but contains less fat and carbohydrate. After 3–4 days milk is produced, but this does not continue unless it is removed. Generally the high concentrations of prolactin produced during lactation suppress ovulation. Oxytocin is not required for formation of milk, but as described in MCQ 130 is important for ejection of milk. FTFTT

MCQ 154

Important in coding the intensity of sensation is

1. the duration of the stimulus.
2. the rate of firing in the afferent nerve.
3. the number of receptors activated.
4. the number of primary afferent fibres activated.
5. the diameter of the afferent fibre.

Apart from the special senses, namely the sensations of sight, hearing, taste and smell, there are many receptors which give information as to the changing external and internal environment—the exteroceptors on the surface of the body, the interoceptors in the viscera, and the proprioceptors in the ligaments, joints, tendons and muscle which give information as to the degree of contraction of the muscles and position of joints in space. The different types of receptors are the mechanoreceptors (e.g. muscle spindles, Pacinian corpuscles), chemoreceptors (e.g. receptors responding to changes in $P\text{CO}_2$), thermoreceptors (cold and warm receptors) and nociceptors (although the latter are probably not specialized receptors). Information within the sensory nervous system can only be transmitted by action potentials, which are virtually identical throughout the nervous system. The specificity of the receptors, the afferent pathways and the brain areas they supply allow the type of sensation and its location to be identified. Thus stimuli transduced in the retinal receptors are interpreted as being due to light so that any stimulus, including a blow, results in the impression that light has been seen—hence the expression 'seeing stars'. The diameter of the afferent fibre is related to the modality rather than the intensity of the sensation. Thus pain and temperature are carried in small myelinated and unmyelinated fibres and light through large myelinated fibres. The intensity of stimulation is reflected by the frequency of action potentials and the number of receptors and hence fibres responding. Increasing the duration of the stimulus will either give rise to the same sensation or the perception of sensation may merely disappear. This is associated with a decrease in the frequency of action potentials in the afferent nerve in the face of a continuous stimulus and is called adaptation. FTTTF

MCQ 155

The receptors which may respond to forms of mechanical deformation are

1. hair follicle endings.
2. aortic baroreceptors.
3. Ruffini's corpuscles.
4. Pacinian corpuscles.
5. free nerve endings.

There are in the skin three main types of sensory receptors: mechanoreceptors, which respond to light mechanical stimulation; thermoreceptors, responding to changes in temperature in the environment; and nociceptors, sensitive to damaging mechanical or thermal stimuli. Other receptors responding to mechanical deformation include baroreceptors, muscle spindles, Golgi tendon organs, labyrinthine hair cells and cochlear hair cells. In the skin the receptors may be either encapsulated or free nerve endings. The anatomical sensory receptors and the probable stimuli to which they respond are as follows:

Sensory receptor	*Probable cutaneous sensation*
Hair follicles	Touch
Naked nerve endings	Pain, temperature, touch
Expanded ends of sensory fibres	
Merkel's discs	Pressure
Ruffini's corpuscles	Pressure, warmth
Encapsulated endings	
Krause's end bulbs	Touch, cold
Meissner's corpuscles	Touch
Pacinian corpuscles	Touch pressure, high-frequency vibration

A group of sensory receptors are served by one sensory nerve fibre. The area of the sensory periphery that affects discharge in a sensory nerve is called the receptive field of that nerve. Receptive fibres from adjacent neurones overlap. The ability to recognize the closely applied stimuli as separate depends on the number of touch receptors, the size of the receptive field and the extent to which they overlap. Spatial discrimination is high over fingers and poor over most of the trunk. There are two types of thermoreceptor: warm and cold. Cold points are much more numerous than warm points. TTTTT

MCQ 156

Generator potentials are similar to synaptic potentials in that they

1. are all-or-nothing events.
2. are always hyperpolarizing.
3. may be summated.
4. are graded in amplitude.
5. are passively conducted.

When energy falls on a receptor cell, be it electromagnetic, mechanical or chemical, it is converted into electrical energy. Specificity is seen because not all kinds of energy bring about the required permeability changes—there is effectively a specialized energy filter. This process of transduction is also one of amplification, so that for example a single quantum of light can trigger membrane changes which influence the visual cell. Like all excitable cells, receptor cells have a resting potential and with an adequate stimulus there is usually a change in membrane conductance to sodium ions, producing a local depolarization termed a receptor potential. This then evokes a generator potential (with similar properties) in the sensory nerve terminal, either directly by electronic spread or indirectly by the release of chemical transmitter. If the stimulus acts directly on a nerve ending, the receptor potential and generator potential are the same. Although depolarization is the rule, there are exceptions; thus retinal rods and cones are hyperpolarized by light and vestibular hair cells may be depolarized or hyperpolarized, depending on the direction of displacement. Depolarization of the receptor membrane lasts as long as the stimulus lasts, the rate of change of potential being proportional to the rate of application of the stimulus. The amplitude of the generator potential is graded, its size increasing with stimulus intensity, so it is not an all-or-none response. The fact that receptors and generator potentials are non-propagating and graded means that they are similar to synaptic potentials. Another point of similarity is that repeated stimuli can summate. The first action potential generated during a stimulus produces a hyperpolarizing after-potential, during which phase the fast sodium channels in the nerve membrane can recover, enabling a maintained stimulus to produce a rhythmic series of action potentials. Phasic receptors adapt rapidly and are responsive to a change in the environment. Tonic receptors adapt slowly, giving constant information as to the relationship of the body to the surroundings. A final group of fibres respond initially with phasic onset—one or two action potentials followed by inactivity. FFTTT

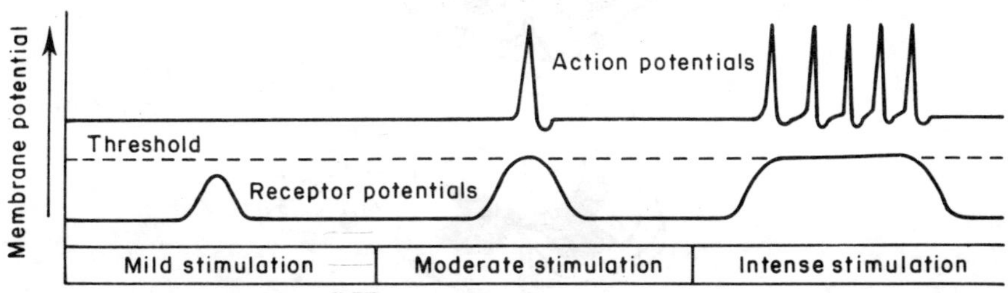

Figure Diagram showing relationship between intensity of stimulation, receptor potential and action potentials.

MCQ 157

The spinothalamic tracts in man

1. comprise axons with cell bodies in the contralateral dorsal horn grey matter.
2. carry sensory information from the thermoreceptors and nociceptors.
3. have the representation from the arm in a lateral position and from the leg in a medial position at C_2.
4. synapse with third-order neurones in the ventroposterolateral nuclei of the ipsilateral thalamus.
5. have axons which decussate within the spinal cord.

A cross-section through the spinal cord reveals a central H-shaped grey area containing nerve cells surrounded by a white area occupied by nerve fibres. The

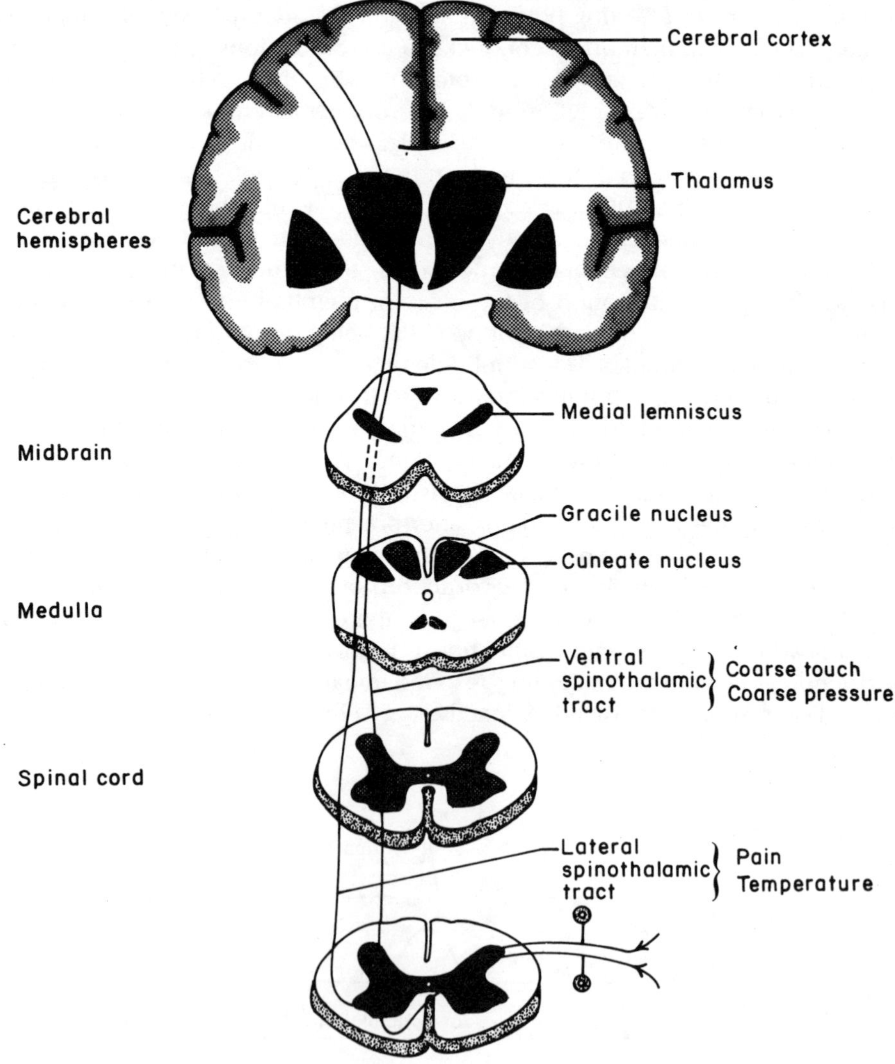

Figure Diagram showing spinothalamic pathways.

fibres travelling through the spinal cord are located in the columns termed dorsal, lateral and ventral, according to their positions. They are also named in terms of their origin and destination. Thus the spinocerebellar tracts run from the spinal cord to the cerebellum. The courses of the pathways have been determined by noting the results of nerve degeneration. The pathway of the degenerated tract may be traced because of the alteration in staining properties. The spinothalamic tract carries information concerning the gross localization of temperature and pain. Sensory nerves enter the spinal cord via the posterior or dorsal nerve roots, the cells of origin lying outside the spinal cord in the spinal ganglia. Fibres leading to the spinothalamic tract relay shortly after entering the dorsal root ganglion. The cells in the dorsal horn are thus second-order neurones. The majority of axons from these neurones cross to the opposite side in the spinal cord and run up in the ventral (anterior) and ventrolateral (anterolateral) white columns. There is a relay in the hypothalamus and the third-order neurones relay impulses to the sensory cortex. Information from the proprioceptors passes up the dorsal columns. These fibres terminate in the cuneatus or gracile nuclei in the medulla. The second-order neurones originating in these nuclei cross the midline and travel in the medial lemniscus to the thalamus of the opposite side. As additional fibres are added to the dorsal columns those from the feet are pushed medially. However, since the spinothalamic tract crosses, the fibres from the lower extremities pass laterally and those from the upper extremities medially. TTFTT

MCQ 158

In the refractive system of the eye

1. greater refraction is produced by the cornea than by the lens.
2. less refraction occurs at the outer face of the cornea as compared with the inner face.
3. in the young adult the lens can double its refractive power.
4. the back surface of the lens is more effective in contributing to accommodation.
5. the ciliary muscles play a role in controlling the refraction of the eye.

The eye is almost spherical in shape apart from the cornea, which bulges outwards. It is enclosed in three layers. The outer layer or external tunica comprises the tough opaque sclera and the transparent cornea at the front of the eye. The next layer—the middle tunica—comprises the choroid and at the front the iris. It also contains the smooth muscle fibres which regulate the curvature of the lens lying behind the cornea and thus also influences the refraction of light by the lens. This allows accommodation so that the eyes can focus on objects from a few centimetres away to the far distance. The internal tunica has two layers: the pigment epithelium and the light-sensitive retina. The eye in many ways is like a camera. The light enters through the cornea and by means of a refractory system, which includes a convex lens, produces an inverted image on the light-sensitive retina. Before the light can reach the lens it passes through the cornea, aqueous humour, lens and vitreous humour. Light is refracted at the highly curved cornea and at the lens. The resting eye has a power of 67 dioptres, with a power of 45 dioptres coming from the cornea. This unit of power, dioptres, represents 1/focal length (in metres). The focal length represents the distance before parallel rays passing through the lens come to a focus. The retina should be at the focal length of the lens, but sometimes the eye may be too short (hypermetropia) in which case the person is long-sighted, or too long (myopia) in which case the person is short-sighted. The refractive defects may be corrected by the use of additional lenses, convex for hypermetropia and concave for myopia. The power of the lens required may be easily calculated as the combined power of two or more lenses used together is given by algebraic summation. The loss of accommodation on ageing, when the near-point (the closest distance at which an object can be clearly seen) recedes, is called presbyopia. TFFFT

MCQ 159

The pupils of man are normally

1. the same size in the light as in the dark.
2. the same size in both eyes.
3. smaller when looking at a near object as compared with a far object.
4. dilate on application of adrenaline.
5. dilate on application of anticholinesterase.

Normally both pupils are round and the same size. Their diameter can be varied by contraction of the muscle systems of the iris. Contraction of the annular sphincter muscle causes constriction of the pupil—miosis,—while contraction of the dilator muscle increases the size—mydriasis. Size decreases with advancing age. It can be altered by the degree of illumination. Light on the eye causes constriction of the pupils. This light reflex involves a pathway from the retina to: (1) the pretectal nuclei, just rostral to the superior colliculi; (2) the Edinger-Westphal nuclei; and (3) the ciliary ganglion and ciliary nerve. When light is shone into one eye the pupils of both eyes constrict. During accommodation to view a close object the pupils are constricted to increase the depth of focus and so improve the image on the retina. In contrast, general activation of the sympathetic system as in the fight or flight response leads to pupillary dilatation. In general adrenergic nerve impulses and noradrenaline produce an increase in the size of the pupils and cholinergic nerve stimulation and parasympathomimetics such as pilocarpine produce pupillary constriction. Cholinergic receptor blockers such as atropine produce pupil dilatation. The diameter of the pupils and the pupillary reflexes are important diagnostically as they can indicate lesions in the oculomotor region of the brain stem and other regions as well as the retina and optic nerve. The pupillary light reflex is one of the last reflexes to disappear in deep anaesthesia, along with the respiratory and cardiovascular reflex control mechanisms. FTTTF

MCQ 160

The cones of the retina in comparison with the rods

1. are more densely packed in the fovea.
2. occur in greater numbers in the human eye than in most other species.
3. function at low light intensity.
4. are more sensitive to longer wavelengths.
5. make approximately 1 : 1 connections with ganglion cells via the bipolar cells.

The light-sensitive retina is made of a mosaic of photoreceptors comprising rods and cones. Before reaching the rods and the cones the light must pass through the vitreous humour, nerve fibres, ganglion cells and bipolar cells. The rods are responsible for twilight (scotopic) vision. They are sensitive to very small amounts of light, but only allow perception of shades of grey, allowing distinction of degrees of brightness, but not colour. Rods are more numerous than cones in the periphery of the retina and do not occur in the fovea. Thus as soon as an object such as a star is fixed visually at night it disappears. The pigment in the rods is rhodopsin, derived from vitamin A, with a maximum absorption at 507 nm. This photosensitive pigment comprises a glycoprotein, opsin, and a pigment, retinol, which is a derivative of vitamin A. The multi-stage breakdown of rhodopsin to transretinol and opsin results in the production of a receptor potential. Cones, which are most numerous in the centre of the retina, are needed for colour vision. The cones have been found to contain pigments which are blue-sensitive (cyanolabe, with a peak sensitivity of 445 nm), red-sensitive (erythrolabe) or green-sensitive (chlorolabe). This was predicted by the Young–Helmholtz theory of vision or the three-colour theory. White light and hence effectively the sensation of all colours may be produced from lights of three primary wavelengths: red, green and blue. Failure of one of the types of cone leads to one type of colour-blindness in which only two or three colours are seen. In the central fovea each receptor cell is linked with a bipolar cell, which in turn is connected to a single ganglion cell, the axon of which joins the optic nerve. In the rest of the retina several sensory cells are linked to a bipolar cell and several of these are linked to a single ganglion cell. Bipolar cells are involved in the process of light–dark adaptation. Horizontal and amacrine cells modulate and transform information conveyed to the brain. The eye is able to work over a wide range of light intensities but adaptation is required, for example, when one goes from a bright room into the dark. Adaptation may take a little time, about 8 min for cones and 30 min for rods. This is brought about by changes in sensitivity or operating range for the level of light intensity of the receptor cells and bipolar cells. TFFFF

MCQ 161

The visual acuity of the human eye

1. is the smallest distance by which two lines may be separated visually.
2. is commonly measured clinically by the use of Snellen's test charts.
3. can be reduced by opacities in the vitreous lens.
4. is independent of the level of illumination.
5. is greater when one eye is used.

Visual acuity is the smallest distance by which two parallel lines may be separated. It is expressed in terms of the reciprocal of the angle the points subtend at the eye. It is tested by using printed letters of such a size that they subtend an angle of one minute if viewed from a specified distance. Clinically the Snellen test is the one most often used. Visual acuity depends on the fidelity with which the light rays are transmitted through the eye, on the density of the retinal receptors and on their innervation. The visual acuity varies over the retina, being greatest in the central region, especially in the fovea. In this region the receptors are densely packed and the ratio of receptors to ganglion cells is 1:1. In the periphery visual acuity is poor because there is much convergence in the rod pathways. However, this anatomical arrangement does allow an action potential to be produced in ganglion cells with very low levels of light. Visual acuity and the contrast between an object and its background may be enhanced by processes within the retina. Like bipolar cells most ganglion cells have receptive fields consisting of a central region with an antagonistic surround, many of them having in addition the property of responding only transiently to changes in retinal illumination. Sometimes a burst is found when there is an increase in illumination (on-response), sometimes with a decrease (off-response) and sometimes there is a burst at the beginning and end of the illumination (on–off response). Normally the eye is not fixed on an object but scans the area, an action which helps visual acuity. The image fades if the movements are blocked and it appears that the on–off response may be important in this respect. Another important property of the eye is that if light stimuli follow each other rapidly they can fuse; the lowest frequency at which this happens is called the fusion frequency. The fusion frequency depends on the level of illumination, the wavelength of the light and the size of the illuminated surface. Thus 20 frames per second may give a flicker-free impression, but 60 frames per second may be required in daylight. TTTFF

MCQ 162

Typically the result of damage to the visual pathway in

1. the right optic nerve is blindness in the right eye.
2. the right optic tract is a homonymous field defect in the left visual field of both eyes.
3. the crossed fibres in the optic chiasma from the nasal portion of the retina is the loss of temporal field vision.
4. the occipital lobe is loss of central vision.
5. optic radiation is loss of visual field on the same side as the damage.

The image formed on the retina is inverted and reversed from right to left, but processing allows for this. Important in the arrangement of the visual pathway is the fact that the portions of the visual field of each eye that overlap are projected onto the same area of the cortex. Lesions of the visual pathway produce characteristic visual field defects, so that the site of the lesion may be determined. About one million fibres arise from each retina and these bundled axons of the large ganglion cells form the two optic nerves. Severe damage to one of the optic nerves leads to blindness in the eye on the same side. After passing through the optic foramen the optic nerves come together to form the optic chiasma, within which partial decussation occurs. Fibres from the nasal half of each retina cross to the opposite side, whereas those from the temporal halves of the retina stay on the same side. A lesion of the optic chiasma thus influences the fibres from the nasal portion of one eye that cross to join the fibres from the temporal half of the other eye. This leads to blindness in one half of the visual field, or hemianopia—in this case bitemporal hemianopia. A lesion of the optic

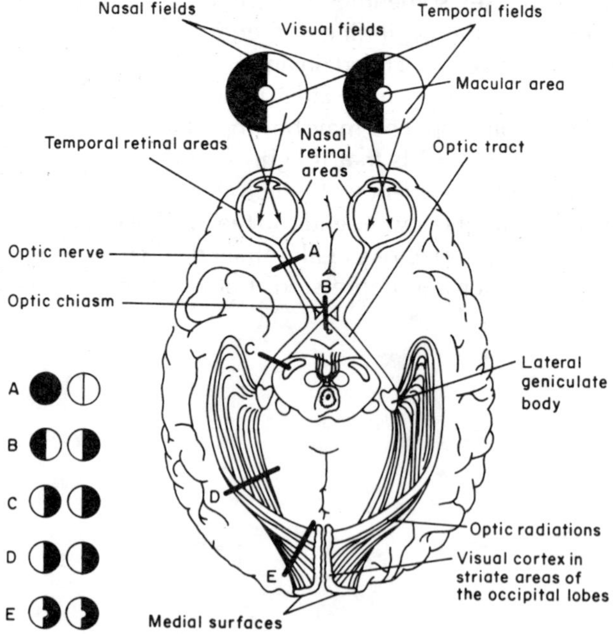

Figure A diagram of the visual pathways indicating the visual field defects resulting from lesions in (A) the optic nerve, (B) the optic chiasma, (C) the optic tract, (D) the optic radiation and (E) the occipital cortex. The disturbances of the visual field are shown on the left.

tract interrupts pathways from the same visual fields in each retina, so that homonymous hemianopia occurs. The term homonymous indicates that the same side of the visual field is affected in each eye. The fibres in the optic tract synapse in the lateral geniculate body of the thalamus. Fibres pass from there to the primary visual cortex via the geniculocortical tract or optic radiations. Lesions here have a similar effect to those in the optic tract. In the visual cortex the area associated with the macula or fovea is greater than the rest of the field. With lesion of the occipital lobe there is some sparing of macular vision. Some fibres from visual pathways transmit signals to the midbrain, so co-ordinating eye movements and pupillary reflexes. TTTFF

MCQ 163

In man the middle ear

1. normally contains air at atmospheric pressure.
2. is connected to the pharynx via the Eustachian tube.
3. ossicles transmit vibration from the tympanic membrane to the oval window membrane.
4. ossicles reduce the intensity of sound reaching the organ of Corti at normal sound levels.
5. ossicles are usually damaged in conductive deafness.

The ear is an extremely sensitive organ, responding to incredibly small energy inputs. Hearing is basically the interpretation of sensations produced by vibrations corresponding to a given pitch, timbre and loudness. The ear has three parts: first, the external ear, which consists of the pinna, the external auditory meatus and the tympanic membrane; next, the air-filled middle ear, which contains the auditory ossicles; and finally the inner ear, where the sound is detected and which also provides sensory input for the maintenance of postural equilibrium. The primary function of the middle ear is to convert the high amplitude, but low force of movement of air into movement of the fluid, with high inertia, of the inner ear. This is achieved by reducing the area over which the force acts. The outer tympanic membrane has an area of 0.7 cm^2 as opposed to the 0.03 cm^2 of the oval window. The lever action of the ossicular bones also contributes to the effect. The ossicles are the malleus, incus and stapes, which are linked together in sequence by ligaments. The arrangement of the ossicles also protects the inner ear from excessive movements. Contraction of the tensor tympanic and stapedius muscles also prevents too great a movement of the oval window. The arrangement of the ossicles also decreases bone conduction of sound from the larynx, so that one is not deafened on speaking. The middle ear is connected to the pharynx via the Eustachian tubes, which act to equalize pressures. Impaired transmission of sound in the middle ear results in conduction deafness. It can occur in chronic otitis media, when the ossicles may be damaged, and in otosclerosis, when the attachments of the footplate of the stapes of the oval window become abnormally rigid. Sensory deafness, in contrast, is due to disease of the cochlea or the auditory nerve and its nuclei. TTTTT

MCQ 164

The basilar membrane of the ear

1. is composed of separate and independently acting transverse fibres.
2. is broader at the base of the cochlea than at the apex.
3. vibrates maximally at the apex in response to vibrations of low frequency.
4. when it vibrates generates the cochlear microphonic.
5. vibrates with greater amplitude the louder the sound.

The organ of hearing is the cochlea, which is located in the inner ear. It is a tube which is coiled through 2½ turns round a central bony pillar, the mediolus. The tube itself comprises three parallel canals: the inner scala media, filled with endolymph; and outer scala tympani and scala vestibula, which communicate with each other at the top of the coil and are filled with perilymph. Reissner's membrane separates the scala vestibula from the scala media and the basement membrane the scala tympani from the scala media. The movement of the oval window induced by the auditory ossicles produces motion of fluid in the scala vestibula, which causes movement of the basilar membrane. Finally the pressure changes are transmitted via the scala tympani to the round window, lying immediately below the oval window. The movement of the basilar membrane stimulates the organ of Corti—the sensory elements in the inner ear—which results in the transmission of impulses via the auditory nerve and several relays to the auditory cortex. The neural impulses are triggered by the development of receptor potentials which may be detected outside the ear as the cochlear microphonic. The basilar membrane is composed of fibres which are 0.15 mm long at the base of the cochlea just outside the oval window. Here the membrane is relatively stiff and high frequencies tend to produce the maximum effect. At the apex the fibres are 0.4 mm long and the membrane is less stiff and responds to low frequencies. The point in the basilar membrane at which the maximum effect occurs is characteristic of a given frequency which activates hair cells and nerves at this point, allowing recognition of frequency. FFTTT

MCQ 164

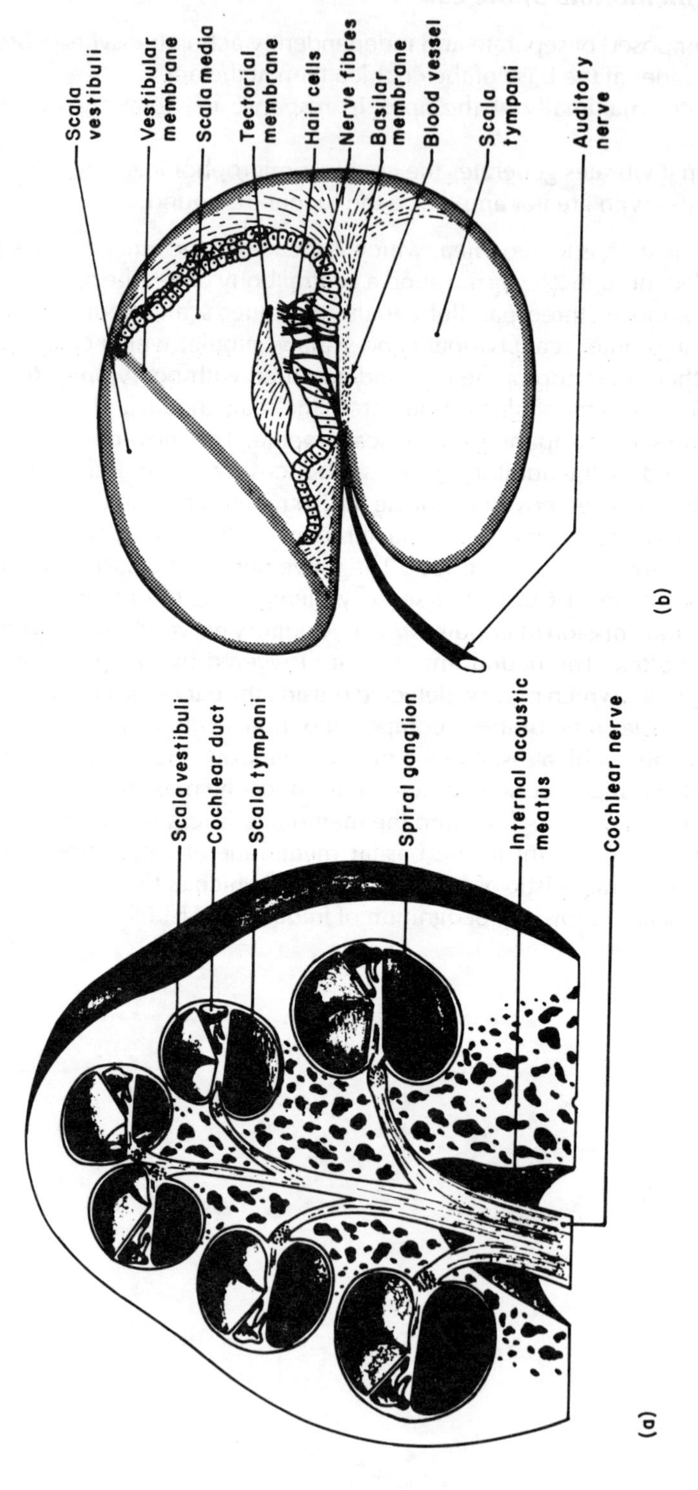

Figure (a) Section through the spiral of the cochlea (after Rohen). (b) The hair cells of the organ of Corti and its relation to the basement membrane.

MCQ 165

When a person is rapidly turned to the right

1. the cupula of the right lateral semicircular canal is deflected away from the utricle (located in the midline).
2. the ciliary processes of the hair cells in the cristae are bent away from the direction of movement in the endolymph.
3. there is a decrease in the activity of the afferent neurones of the right semicircular canal.
4. the eyes are caused to move slowly (track) to the right.
5. the direction of nystagmus (fast phase of eye movement) is to the right.

In addition to the sense of hearing the ear provides information concerning gravity, rotation and acceleration, which is necessary for the maintenance of balance and equilibrium. This involves the vestibular apparatus, which consists of the saccule and utricle and the three semicircular canals. The saccule and utricle respond to the position of the head in space and to linear acceleration, including gravity. They contain maculae, epithelial cells containing calcified particles called otoliths, which represent an inertial mass which tends to remain stationary as the head moves, distorting the cilia and thereby initiating action potentials. The three semicircular canals lie at right angles to each other, so that acceleration is determined in any plane. At one end of each canal there is an enlargement—the ampulla—containing specialized epithelium located in a transverse ridge and known as cristae. During rotational movement the endolymph of the semicircular canals moves and distorts the cristae, stimulating hair cells. The cristae are thus in some ways analogous to the maculae, but have the same density as endolymph. Information from the vestibular organs is transmitted to the central nervous system via the vestibular division of the cranial nerve. Numerous connections exist between the vestibular nuclei and the cerebellum and other regions, which allows integration of vestibular impulses with other sensory functions concerning balance and equilibrium. On turning the head rapidly there is an eye movement against the direction of movement, so that the original direction of gaze is maintained. Before they reach the limit of their range the eyes are flicked back in the direction of rotation. The shifting movement of the eyes is known as nystagmus; the direction of nystagmus is usually taken as the direction of the fast phase. When the head is rotated to the right, the fluid and the hairs move to the left and in the case of the right ear towards the utricle with an increase in the afferent discharge. TFTFT

MCQ 166

Odours

1. contribute to the sense of taste.
2. are detected by specialized nerve endings.
3. are less readily detected on ageing.
4. are adapted to within the first few seconds of exposure.
5. may be more readily detected at certain stages of the menstrual cycle.

Odours are detected by olfactory cells, which are nerve cells with hairs projecting from the surface of the epithelium and are embedded in a layer of mucus. The primary afferent neurones are separated from each other by supporting cells, which may have a metabolic function. Odour molecules must be volatile and water-soluble to allow them to dissolve in the mucus and also lipid-soluble to reach the receptor. They are conveyed to the receptors on inspiration or from the oral cavity. Sniffing may aid in the detection of odours by directing air currents across the nasal mucosa and thus increasing the concentration of active molecules in the mucosa. Man can distinguish somewhere in the region of 3000 different odours, but attempts to classify them have met with limited success. Six basic odour classes have been suggested: floral, ethereal, musky, camphor, putrid and pungent. The basis for odour discrimination is little understood. It is thought that the molecular shape is of key importance, as it is possible to distinguish between isomers of the same substance. The axons from the olfactory receptor cells are fine myelinated fibres which run in bundles or fascicles. They pass to the olfactory bulb and enter the glomeruli, where they synapse with dendrites from mitral and tufted cells. These cells form the second-order neurones in the olfactory pathways. Axons from mitral cells run to the olfactory cortex. Other fibres connect with the limbic system, which may be why there is a marked emotional response associated with certain odours. The sense of smell shows adaptation, so that one becomes aware of odours be they pleasant or unpleasant; however, this does not occur extremely rapidly. As with other special senses there may be impairment on ageing. There may be differences in the olfactory threshold for some macrocyclic molecules (related to pheromones) over the menstrual cycle. TFTFT

MCQ 167

Current concepts of pain hold that

1. the sensation of pain results from activation of small-diameter fibres.
2. the intensity of pain can be reduced by activity in the large afferent fibres from the region activating the small fibres.
3. physical stimuli producing pain may act via chemical mediators.
4. no specialized receptor cells exist for pain.
5. activity in the A fibres produces a pricking pain.

Pain may be described as the sensation resulting from stimuli which are intense enough to threaten or cause injury. Pain protects against tissue damage; those without the ability to feel pain suffer from infections and pathological changes in the spine, bone and joints. Pain can be produced by many kinds of physical stimuli: thermal, mechanical and electrical. It is possible that physical stimuli act via chemical mediators, including kinins, prostaglandins, hydrogen ions, potassium ions, histamine and acetylcholine. It used to be thought that there were no specific pain receptors, the sensation of pain being produced by overstimulation of receptors sensing other sensory modalities. It appears that pain or nociceptors are probably free nerve endings. They are supplied by the small myelinated A fibres 2–5 μm in diameter and the unmyelinated C fibres 0.4–1.2 μm. Impulses in the former, which are conducted at a rate of 12–30 m s^{-1}, signal high-intensity mechanical stimulation, giving rise to a sharp localized sensation. The C fibres, in which impulses travel at 0.5–2 m s^{-1}, are less selective and their stimulation gives rise to a dull ache. There are many different types of pain, but they may be classified according to the site of origin—visceral pain, somatic pain, which may be further subdivided into cutaneous pain, and deep pain coming from the muscles, joints, bones and connective tissue. Because the viscera contain fewer receptors visceral pain is poorly localized and may be referred to other body areas. Pain sensation is not directly related to the extent of the injury, but depends on other factors such as the location of the damage, condition of the subject, and possibly race and culture. Schemes have been put forward to describe the neural activity underlying the various forms of pain, including the gate theory of Melzak and Wall, which postulates that the substantia gelatinosa gates activity from the pain fibres. Inhibition can occur at this site, where enkephalins—endogenous opioid peptides—are found. The activity of larger input fibres may suppress pain impulses from smaller pain fibres. TTTTT

MCQ 168

Referred pain

1. appears to originate in visceral structures.
2. may appear to come from the left shoulder and upper arm if the source of pain is the heart.
3. may be blocked if local anaesthetic is applied to the area to which the pain is referred.
4. has no clinical significance.
5. arises in an area sharing the same segmental innervation as the organ causing the pain.

An important characteristic of pain relative to central nervous processes is its tendency to irradiate and to give rise to referred pain. Thus damage to an internal organ may be associated with pain or tenderness, not in the organ, but in a region sharing the same innervation. For example, cardiac pain may be referred to the inside of the left arm and that from the central portion of the diaphragm to the tip of the shoulder. The probable explanation is that some central cells receive both cutaneous and visceral inputs. Such cells have been described in lamina V of the dorsal horn. Normally such cells are activated by skin inputs. When the visceral organ is injured it activates the same cell in the dorsal horn and this is interpreted as skin injury. The skin feels tender and local anaesthesia apparently helps to relieve the pain by reducing the total sensory input from the skin. Another type of inappropriate pain is phantom limb pain. Following amputation some patients are left with persistent pain apparently originating from the amputated limb. The pain is believed to arise centrally. Temporary analgesia only is produced by section of the contralateral and anterolateral tracts. An important neural system for inhibition of pain is that which arises from the mesencephalic grey matter and descends to dorsal horn cells in the spinal cord. The system can be activated by electrical stimulation to produce marked analgesia; it contains opiate receptors to which morphine becomes bound before it produces analgesia. It has been suggested that endogenous opioid peptides which bind to the same receptors as opiates may play a role in acupuncture, as it has been claimed that the effects can be blocked by naloxone (an opioid receptor antagonist). Analgesia may be produced by local anaesthetics which block conduction in the periphery by raising the threshold of the neurones. Central anaesthetics are thought to act by interfering with transmission of neural activity in the reticular formation. FTTFT

MCQ 169

A motor unit

1. comprises a motor neurone and the muscle fibres it innervates.
2. has a motor neurone which leaves the spinal cord via the dorsal horn.
3. when active produces fibrillation.
4. maintains muscle tone by acting with other muscle units.
5. has fewer muscle fibres linked to the motor neurone in the extraocular eye muscles than in the quadriceps.

The motor system comprises all the structures controlling the activity of muscle. Control of motor activity is via descending pathways, the majority of which appear to end on an interneurone in the reflex arc rather than directly. There are five important descending tracts (see also MCQ 176). The corticospinal or pyramidal tract contains about 3 per cent of fibres which make direct connections with the motor units. The other four pathways come from closely neighbouring parts of the brain, the brain stem and medulla, the reticular formation, the vestibular nuclei, red nucleus and tectum. The final common pathway comprises the motor unit, made up of a motor neurone, the axon of which leaves the spinal cord in the ventral root, and the muscle fibres it innervates. The number of fibres supplied by the nerve can be 500–2000 in a powerful limb muscle. Fine control of movement is achieved when the fibres in a motor unit are reduced to five to ten, as in the eye muscles. The fibres supplied by one neurone are not necessarily close to one another, but may be distributed throughout the muscle so that a balanced contraction of the entire muscle may be obtained. The timing and force of contraction depends on the pattern of the action potentials. This can also occur by an increase in the rate of firing in one motor unit. Within the body rapid twitch contractions are not seen; instead twitch contractions are summed to produce a state of constant or fused contraction called tetanus. This can occur since contraction of skeletal muscle lasts longer than its membrane refractory period, so that it is possible to produce a second contraction before relaxation is completed. Fibrillation is the unco-ordinated contraction of muscle fibres due to muscle disease and is also seen in the heart. Alpha motor neurones are multipolar cells (many dendrites but one axon), the dendrites spreading from the ventral horn to the dorsal horn and even across then up and down a given segment of the cord, allowing extensive synaptic contact and the convergence of many inputs into the final common pathway. The cell bodies of motor neurones of the trunk and proximal muscles lie in the medial part of the ventral horn. The more lateral nuclei belong to the more distal muscles. TFFTF

MCQ 170

Contraction of the intrafusal muscle fibres

1. stimulates the flower spray endings.
2. may be produced by activation of the γ efferent neurones.
3. on section of the dorsal roots produces a large increase in the tension of the extrafusal muscle.
4. reflexly results in relaxation of the muscle.
5. increases the sensitivity of the stretch reflex.

Mechanoreceptors are found in muscle which provide information about its length and tension and thus contribute to the control of posture in addition to providing conscious sensation. These receptors are associated with muscle spindles and with tendon receptors called Golgi tendon organs. The muscle spindle is located within the muscle in parallel with the other muscle fibres, many spindles being found in each muscle. The muscle spindle is made up of two to twelve small-diameter muscle fibres termed intrafusal fibres, in contrast to the extrafusal fibres of the main muscle. The outer thirds of the intrafusal fibres are striated and contractile, whereas the middle third, the equatorial region, possesses no striations. Two types of intrafusal muscle fibres exist. First are the nuclear bag fibres, of which there are generally two per spindle and which are larger and thicker and have a group of nuclei clustered in the centre; second are the nuclear chain fibres, of which there are four to five per spindle and which are shorter and thinner. Located in the equatorial region of these fibres are mechanoreceptors which are stimulated when the spindle is stretched, as will occur when the whole muscle is stretched and pulls on the end of the spindle. Two types of receptor are found: the annular spiral endings in nuclear bag fibres which form a primary receptor ending, supplied by group Ia fibres; and flower spray endings in the nuclear chain fibres which form secondary receptor endings supplied by group I fibres. Both types of nerve ending give a static response, their rate of firing being proportional to muscle length. Spindles also show a dynamic response, the firing rate depending on the rate of change of length of the muscle. Muscle spindles also have their own motor innervation via the fusimotor or γ motor neurones. This produces contraction of the polar region of the intrafusal fibres, stretching the equatorial region and altering the sensitivity of the muscle spindle. When the muscle is passively stretched, the muscle spindle is stretched, activating Ia endings which in turn results in a reflex contraction of the extrafusal muscle fibres to return them to their original length. TTFFT

MCQ 171

In a stretch reflex

1. a single synapse is involved.
2. impulses travel via Ia muscle afferents.
3. the sensory ending involved is the Golgi tendon organ.
4. the antagonistic muscles relax.
5. the reflex response time is about 100 ms.

A reflex action may be defined as an involuntary response to a stimulus. It is thus a stereotyped response over which little voluntary control is exerted, although it is possible. A reflex involves at least five components: a sensory receptor, an effector and at least three neurones, a sensory neurone, an interneurone and a motor neurone. The cell body of the sensory neurone lies in the dorsal root ganglion situated just outside the spinal cord. The neurone is bipolar, the long axon extending to the receptor while the shorter enters the dorsal root. Sensory neurones may connect with interneurones or directly to the motor neurone. In the stretch reflex only two neurones are involved in the reflex arc, which is unique to the stretch reflex. This reflex is seen, for example, when the tendon below the patella is tapped to elicit a quick jerk of the leg. The stretch of the muscle leads to a discharge in the muscle spindle afferents, resulting in contraction of the α motor neurones of the same muscle. Each synapse crossed in the reflex adds to the delay of transmission, so that the monosynaptic pathway, which also contains Ia fibres which are large and fast, has the least delay, of the order of 30 ms. The reflex nature of the response is shown by the abolition of the response on cutting the dorsal roots. The reflexes, however, persist after all cutaneous nerves have been cut and the tendon organs anaesthetized. Stretch reflexes are present to some extent in all muscles, but are most noticeably seen in the antigravity muscles. They are susceptible to inhibition. TTFTF

MCQ 172

The γ loop has great importance as it contributes to

1. the smoothing out of muscle movements.
2. maintenance of muscle tone.
3. the regulation of phasic stretch reflexes.
4. postural adjustments.
5. the initiation of movement.

As described in MCQ 169 and 170 there exist receptors in muscles called muscle spindles, the activity of the efferent nerves of which affects the degree of muscle contraction, which in turn can influence muscle length via the stretch or myotatic reflex. The length of the muscle spindles themselves can be influenced via the γ motor neurones. This is important as it allows the muscle spindles to work over a wide range of muscle lengths and also means that both α and γ motor neurones produce movement. Important in muscle contraction is the γ loop. It is so called because it comprises the γ efferent fibres sending impulses to the muscle spindles and the afferent fibres sending information back to the central nervous system to influence activity in α motor neurones. Activity of γ efferents is controlled by descending pathways. Higher centre control of limb movement usually involves simultaneous co-activation of α and γ motor neurones. Thus if a heavy load is to be lifted and the initial α motor neurone activation and the resulting muscle contraction is insufficient to move the load, then the intrafusal fibres are put under tension and their firing increases the force generated. Muscle spindles of course do not only signal information concerning muscle length, but also the rate of contraction. The loop is essential to preset the muscle to contract to lift a load and to make adjustments in contraction rate and force during the actual process to allow a smooth response. The γ loop is also important in setting muscle tone precisely and hence in the control of posture. Muscle tone is a sustained low-grade reflex contraction of muscle and is necessary as a background for subsequent contraction. It is maintained by the constant pull of gravity in the spindles of the antigravity muscles. Pathologically increased tone, termed spastic tone or spasticity, may result from increased motor neurone discharge. TTTTT

Firing of afferents from the Golgi tendon organ

1. is stimulated by active muscle contraction.
2. is stimulated by passive stretch of the muscle.
3. results in a tendon jerk in the originating muscle.
4. influences motor neurones supplying muscle antagonists.
5. leads to the inhibition of the same muscle.

The Golgi tendon organ, as the name indicates, is found in the tendon usually near the muscle–tendon junction. They are supplied by Ib afferents. Excitation of the Golgi tendon organ occurs in two steps. First, the contraction of muscle fibres inserted in the receptor increases the tensile force of the Golgi tendon organ bundles. This in turn squeezes the afferent terminals interwoven amongst them. The Golgi tendon organ lies in series with the extrafusal fibres and can be activated both by passive and active stretching. However, it is much more difficult to activate the organ by passive stretch, probably because the elastic components of the skeletal muscle absorb much of the tension, i.e. it has a high threshold. Like the muscle spindle it has dynamic and static response characteristics, responding both to the velocity of tension development and to constant maintained tension. Originally it was believed that the tendon organ was important for indicating if excessive stretch or contraction were occurring. However, it has now been shown that it is extremely sensitive, responding to less than 0.1 g force applied directly to the receptor endings. Thus under isometric conditions the organ is directly activated. It does, however, contribute to a protective reflex—inhibition of muscle tension. This is the relaxation of the muscle seen when the tension in the muscle exceeds a certain level. This is particularly marked when muscles are hypertonic. If the leg of an animal with decerebrate rigidity is forcibly flexed it resists the applied flexion until at a certain point the resistance suddenly disappears; this is known as the clasp knife reflex. TTFTT

MCQ 174

Polysynaptic reflexes show

1. irradiation.
2. after-discharge.
3. facilitation.
4. occlusion.
5. temporal interaction.

In contrast to the phasic monosynaptic reflex, the sustained polysynaptic response consists of tetanic contractions with stimulation of both γ and α motor neurones. An example of a polysynaptic reflex is the flexor reflex, a typical example of which is the withdrawal of a limb from a painful stimulus. Nociceptive stimulation can, via a pathway involving interneurones, produce excitation of flexor motor neurones and inhibition of the ipsilateral neurones of the antagonist muscles (reciprocal innervation), as well as excitation of the contralateral motor neurones of extensor muscles. The latter results in the crossed extensor reflex. The degree of involvement of other limbs depends on the strength of the stimulus. Polysynaptic reflexes show irradiation in which the stronger the stimulus the greater the number of fibres involved. Several spinal segments can also be involved. In addition the stronger the stimulus the more prolonged the reflex, as a result of repeated firing of the motor neurones, i.e. the phenomenon of after-discharge. These and other properties depend on the typical arrangement of motor neurones in the central nervous system. Each neurone is the site of convergence for many pathways. Conversely one afferent fibre entering the spinal cord delivers its messages to many pathways. One resulting property is spatial summation. Two afferent fibres when stimulated separately may neither produce a reflex, i.e. each stimulation is subliminal. When the two fibres are stimulated simultaneously there may be summation of the subliminal excitation and a response may be obtained. Of less importance is temporal summation, illustrated by the repetitive stimulation of an afferent fibre evoking a reflex response even though individual stimuli of the same strength would be ineffective. Occlusion is seen, for example, when two afferent excitatory nerves, each of which can evoke the flexor reflex, are stimulated simultaneously and the tension developed is less than the sum of the tensions produced by each afferent stimulated separately. This is due to the fact that there is a motor neuronal pool shared by the group of input fibres that are brought into the firing zone by the individual stimuli. TTTTT

MCQ 175

Acute transection of the lower thoracic spinal cord results in

1. paralysis of the respiratory muscles.
2. spinal shock below the section.
3. loss of all function distal to the block.
4. a rise in blood pressure.
5. immediate incontinence of urine.

The spinal cord runs in a vertebral column from the base of the skull to the upper lumbar region. The cord is segmentally arranged, as witnessed by the 31 pairs of nerves which travel from the cord to the various parts of the body: 8 cervical nerves, 12 thoracic nerves, 5 lumbar nerves, 5 sacral nerves and the coccygeal nerve. Thus there are two distinct patterns of abnormality: segmental, producing signs at a single segment; and longitudinal, producing deficiency of all functions below the level of the lesion. Middle and lower cervical lesions produce segmental signs of motor and sensory dysfunction in the upper extremities, together with long tract signs in the lower extremities and supranuclear bladder dysfunction. The vasomotor activity maintaining arteriolar and venous tone runs down the lateral columns of the cord and leaves via the sympathetic outflow arising from the first thoracic to second lumbar segment. A high thoracic spinal section will interrupt this tone and cause a fall in blood pressure. Section of the spinal cord is followed for several weeks in man by spinal shock, a condition in which all cord functions below the lesion become depressed. There is paralysis, loss of tone and reflexes and loss of sensation. During the period of spinal shock the bladder is atonic and no urine is passed until the bladder overfills, when retention with overflow incontinence thus occurs. In recovery from spinal shock the neurones recover their excitability, a process which in man may take up to several months. The flexion reflexes reappear first. After several months hyperexcitability occurs as may return of muscle tone and tendon reflexes. Thus, although no voluntary movement is possible, reflexes occur. Stimulation below the lesion, as for example by stroking the inner surface of the thighs, may give rise to a mass reflex with flexor spasm of the legs, instant evacuation of the bladder and sweating below the level of the lesion. FTTFF

MCQ 176

Damage which interrupts upper motor neurone pathways in the human would lead to

1. voluntary muscle paralysis.
2. exaggerated tendon jerks in antigravity muscles.
3. spasticity of limb muscles.
4. wasting of muscle.
5. fasciculation.

Lesion of the motor pathways from the cerebral hemispheres are quite common as a result of cerebrovascular accidents and are called upper motor lesions. Specific lesions of the direct corticospinal pathway rarely occur, but rather damage to both pyramidal and extrapyramidal pathways as they are intermingled. The pattern of loss of function depends on the site of the lesion. Because fibres cross in the medulla lesions about this point affect the contralateral side. Also in contrast to the lower motor lesion is the fact that paralysis seen is not accompanied by atrophy and the reflexes are maintained. An upper motor lesion will affect movement, muscle tone and posture and reflexes. Lesions of the corticospinal tract results in loss of fine skilled movement and distal weakness, a decrease in muscle tone, absent abdominal reflexes (reflex contraction of abdominal muscles on stroking overlying skin) and an unwillingness to use the affected limb. On stroking the sole of the foot Babinski's response is seen, in which there is dorsiflexion of the big toe and spreading of the other toes. This is mediated by spinal cord mechanisms akin to a nociceptive withdrawal reflex and in the adult it is normally suppressed by the corticospinal pathway. Gentle stroking of the palm elicits an abnormal grasp response—the grasp reflex. The spastic tone and brisk tendon jerks may be explained in terms of a loss of inhibitory influences which results in an increase in the force of the phasic stretch reflexes, particularly in the antigravity muscles. This may represent an imbalance of influences impinging on the motor neurones as a result of the loss of rubrospinal and medullary reticulospinal pathways as well as the corticospinal pathway. It is damage to the final common pathway which results in fasciculation or single spontaneous discharge of a motor unit. TTTFF

MCQ 177

The control circuits for the co-ordination of postural responses lie in

1. the cerebral cortex.
2. the thalamus.
3. the basal ganglia.
4. the brain stem nuclei.
5. the reticular formation.

Posture represents the maintenance of the position of the head, trunk and limbs as a prelude or background to movement and hence the distribution of muscle tone and the stretch reflex. There are several types of postural reflexes: those evoked by gravitational forces, by linear acceleration and head on body movement (neck reflexes). In man vision is also quite important in the control of posture. The major co-ordinating systems in the maintenance of balance and posture are the cerebellum, the basal ganglia, the brain stem nuclei and the reticular formation. The underlying mechanisms for posture and movement are the same and indeed movement may be necessary to maintain an upright posture. Voluntary control of skeletal muscle by higher centres is via two pathways: the pyramidal or corticospinal pathway and the extrapyramidal pathway. The pyramidal tract, which is one pathway for rapid movement, passes directly from the motor cortex to the decussation within the spinal cord. The fibres cross to the other side in the medulla (pyramidal decussation). The extrapyramidal pathways also originate in the motor cortex and other supra-spinal centres. They consist of the motor cortex, the basal ganglia and several brain stem nuclei, including the red nucleus and olivary nucleus. The function of the basal ganglia is indicated by the effect of damage, namely abnormal movements, increased muscle tone or rigidity and slowness in initiating and changing movement. The most common disorder of the basal ganglia is Parkinson's disease, in which an important loss is that of the dopaminergic pathway to the caudate putamen complex or corpus striatum (other areas of the basal ganglia being the substantia nigra, globus pallidus and subthalamic nucleus). The basal ganglia thus appear to be concerned with the fixing of the body and proximal parts of the limbs to allow a firm base for movement of the hands and feet. The reticular formation is the region for the integration of all influences which are involved in the control of posture. Impulses pass from the reticular formation to the γ motor neurone system. FFTTT

MCQ 178

Necessary components of locomotion include

1. bones to act as a system of levers.
2. antigravity support for the body.
3. control of the centre of gravity.
4. spinal reflexes.
5. an undamaged cerebral cortex.

The maintenance of upright posture in man is a complex affair, presenting more problems than those faced by a quadruped. The skeleton acts as a support and because it is jointed movement between the bones is allowed. They are held together with ligaments and tendons and the muscles are attached to the bones in such a way as to move them as though they were a system of levers. The centre of gravity is in the pelvic region above the small bones of the feet. Small swaying movements occur continuously as the centre of gravity moves and is then restored to normal. It is for this reason that infants take time to learn to stand upright. The mechanisms underlying the corrections entail integration of information from the eyes, vestibular apparatus, the muscles and joints. Thus if muscle joint sense is lost as in tabes dorsalis (destruction of the dorsal roots and dorsal columns of the spinal cord) the patient becomes unsteady, especially if the eyes are closed. There appears to be very early signalling of small deviations and prompt reflex activity, so that, as EMG recordings show, there is relatively little activity in the antigravity muscles, e.g. extensor muscles of the legs on standing. Similar mechanisms are involved in locomotion. Antigravity support is required. Only when the centre of gravity is over one leg can the other be raised. In walking, the centre of gravity is alternately poised over the right and left legs. The basic mechanisms for walking lie in the spinal cord and alternate flexion of one limb and extension of the other can be built up on the crossed extensor reflex, the decision to move coming from the cerebral cortex and co-ordination from the basal ganglia, cerebellum and reticular formation. Stepping can be elicited in newborn infants by holding them so that their feet touch the surface and their legs support their weight. By rocking the child slightly from side to side and tilting them slightly, purposeful stepping may be initiated. Similarly in the adult a slight tilt forward is enough to elicit stepping. The body falls forward and loses its equilibrium until the weight is taken on the leg which was swung forwards. TTTTF

MCQ 179

The cerebellum is involved in

1. the conscious appreciation of limb position.
2. the maintenance of posture.
3. the control of voluntary movement.
4. the performance of rapid alternating movements.
5. and is essential for the stretch reflex.

The cerebellum is known to be concerned with movement because of the disturbances which follow damage to the region, namely disorders of balance and gait, hypotonia and inco-ordination (ataxia). An example of an abnormality is nystagmus, when the eyes drift away from the fixation point towards the centre of gaze and are then moved rapidly back to the fixation position. The cerebellum is a highly convoluted organ dorsally situated over the medulla and pons. The oldest region lies in the midline and is concerned with control of the trunk and limbs. The oldest part is the archicerebellum, associated with the vestibular apparatus and hence important in the control of posture. Information also reaches the palaeocerebellum (phylogenetically an old part) from the proprioceptors, making it important in movement. The newer regions (neocerebellum) push out laterally. The lateral hemispheres have undergone extensive development to allow a large repertoire of finely co-ordinated motor skills, which in the human includes speech. The cerebellum serves a co-ordinating function within the motor system. It does not appear to be involved in conscious sensation, nor does it initiate movement. Instead it continuously compares the output to the muscles with the returning sensory information, so that corrections are made to ensure the desired power and range of motion. To enable this to occur it is linked collaterally with other motor centres and receives information from practically all sense organs, with the exception of smell. There is continuous feedback of information on voluntary and involuntary movements so that fine adjustments can be made even during action. The most conspicuous cell type is the Purkinje cell, which provides the only efferent or output pathway from the cerebellar cortex to the deep nuclei, the effect being purely inhibitory. Afferent information reaches the Purkinje cells via a unique single fibre that climbs up its efferent axon (climbing fibres) and the mossy fibres which synapse with the granule cells, the latter sending out parallel fibres. Each Purkinje cell receives around 200 000 inputs from granule cells. FTTTF

MCQ 180

Characteristic of the organization of the autonomic nervous system is that

1. the cell bodies of the spinal supply lie in the dorsal root ganglion.
2. there are at least two synapses in the outflow pathways outside the spinal cord.
3. all the output pathways originate in the spinal cord.
4. there are always different transmitters at the ganglia and the target organ.
5. sympathetic nerve fibres usually have a synaptic relay close to the target organ.

The autonomic nervous system regulates and co-ordinates the functions of the internal organs and tissues. Its fibres supply cardiac muscle, smooth muscle and secreting glands. Some authorities define the autonomic nervous system as purely motor, although it requires afferent information. Its organization differs from that of the voluntary motor system in that all the final motor neurones lie outside the central nervous system. Thus the efferent pathway comprises two neurones. The first neurone—the preganglionic neurone—which leaves the spinal cord via the ventral horn is slowly conducting. It synapses with the postsynaptic neurone at a ganglion where acetylcholine is the neurotransmitter. The autonomic nervous system comprises sympathetic and parasympathetic divisions which differ in the origins of the preganglionic neurones, the positions of the autonomic ganglia and the nature and effect of the transmitter substances at the target organs. The sympathetic nervous system, in which the post-ganglionic fibres are adrenergic, has a thoracic and lumbar outflow, each preganglionic fibre leaving via a white ramus communicantes and passing to a ganglion. The sympathetic ganglia are linked to form a chain either side of the vertebral column. The fibres supplying the head and breast synapse in two cervical ganglia and upper sympathetic trunk ganglia. The fibres to the abdomen and pelvis pass through the ganglia and synapse in unpaired ganglia. This arrangement means that preganglionic fibres are relatively short and the post-ganglionic fibres long. The reverse is true of the parasympathetic system, in which the ganglia generally lie in the organ supplied. The parasympathetic outflow is cranial (nerves III, VII, IX and X), the vagus (X) being the most important, and sacral. The chemical transmitter at the postganglionic terminals is acetylcholine. In the autonomic nervous system afferent and efferent fibres from sensory receptors run together as visceral nerves. FTFFF

MCQ 181

Changes resulting from parasympathetic stimulation include

1. dilation of the pupil of the eye.
2. slowing of the heart.
3. contraction of bronchial muscle.
4. contraction of the detrusor muscle of the urinary bladder.
5. decreased intestinal motility.

The parasympathetic system innervates the smooth muscle and glands of the gastrointestinal tract, the excretory organs, the genitalia and lungs. It also supplies the atria of the heart and the tear and salivary glands. It does not generally supply smooth muscle of blood vessels except for those of the genitalia and possibly the brain and heart. Generally stimulation of the sympathetic system brings about changes opposite to those of stimulation of the sympathetic system, although the two may act together synergistically, i.e. a response may be brought about by an increase in the activity of one system and a concurrent decrease in that of the other. In some situations mutual antagonism can occur via presynaptic inhibition. Thus the sympathetic supply to the myenteric plexus of the gastrointestinal tract appears to act by presynaptic inhibition of neurotransmitter release. Stimulation of the parasympathetic system brings about its effects which are directed to the maintenance and conservation of body function. The responses include a decrease in heart rate and contractility of the atria, and constriction of the pupils, bronchi, the gall-bladder and detrusor muscle of the bladder. There is increased secretion and motility in the gastrointestinal tract and relaxation of its sphincters. There is also increased secretion from the tear glands. In some organs receiving dual innervation, activity of one system predominates. For example, control of the urinary bladder, diameter of the pupils, of resting heart rate and activity of some exocrine glands depends on the parasympathetic activity. In contrast to the sympathetic system, the parasympathetic system has no effect on metabolism, nor on the muscles of hair follicles or sweat glands. FTTTF

MCQ 182

Increased sympathetic activity can result in

1. an increased rate of depolarization of action potentials in the sinoatrial node.
2. increased gluconeogenesis in the liver.
3. relaxation of the gall-bladder and ducts.
4. diuresis.
5. increases in pancreatic exocrine secretion.

The sympathetic nerves supply the smooth muscle of all organs, namely blood vessels, viscera, excretory organs, lungs, pupils and hair follicles; also the heart and many glands such as the sweat glands, salivary glands and others of the digestive tract. In contrast to the parasympathetic system, fibres of the sympathetic branch also travel to adipose and liver cells and also the adrenal medulla. Stimulation of the sympathetic nerves is involved in the response to stress, the responses as a whole forming the 'flight or fight' response. The changes include an action on the sinoatrial node with an increase in heart rate, an increase in stroke volume and dilation of the bronchi. There is also a decrease in muscle fatigue and an accompanying effect on metabolism, so that energy supplies are increased. This includes glycogenolysis in the liver and lipolysis in fat cells which results in elevated blood glucose and free fatty acid concentrations. There is a marked decrease in gastrointestinal tract activity, with a decrease in intestinal motility, relaxation of the gall-bladder and contraction of the sphincters of the tract, as well as constriction of the splanchnic vascular bed. Blood is also diverted away from the skin. The adrenal medulla is a modified sympathetic ganglion. The cells, which are homologous to postganglionic neurones, are activated via cholinergic synapses with preganglionic axons to release catecholamine. On denervation the autonomic effector organs do not degenerate but may exhibit some atrophy. Depending on the organ involved, supersensitivity develops from 2 to 30 days after denervation. TTTFF

MCQ 183

The autonomic outflow can be altered by

1. afferents from visceral receptors.
2. afferents from somatic receptors.
3. activity in projections from the limbic system.
4. hypothalamic activity.
5. activity in the medulla oblongata.

The main groups of visceral receptors which have been identified are mechanoreceptors and chemoreceptors. These receptors are not only located in the walls of blood vessels, as indicated in MCQ 54 and 75, but also in the viscera of the thorax and abdomen. Mechanoreceptors, for example, give information as to the filling of the gastrointestinal tract and bladder and chemoreceptors on the pH of the stomach contents. Sensations can reach the nervous system directly via the cranial nerves or indirectly via the spinal cord, afferents entering via the dorsal roots. They are then relayed to the brain stem. If there is a change in the autonomic activity produced via the visceral input, the response may be referred to as an autonomic reflex. Control of the visceral systems occurs centrally at several levels: the cortex, hypothalamus and reticular formation. The hypothalamus provides the major control of the reflexes and is known as the head ganglion of the autonomic nervous system. It contains receptors for temperature, osmotic pressure and glucose concentrations in the blood. It receives afferent impulses from the ascending somatic visceral tracts as well as further impulses from the thalamus and limbic system. As described in MCQ 184 the part of the cerebral cortex and underlying structures which are active in visceral and emotional responses are known collectively as the limbic system. In this region experiences are evaluated so that the body is able to modify elementary hypothalamic patterns and adapt them to the prevailing environmental situations. Medullary centres in the midbrain with control of autonomic function are the cardiovascular and respiratory centres. TTTTT

MCQ 184

The ascending reticular formation

1. transmits information to the cerebral cortex.
2. is concerned with the state of consciousness.
3. is activated by collateral branches of sensory neurones conveying information to the thalamus.
4. is concerned with modulation of sensation.
5. can excite the limbic system.

The reticular formation is an elongated structure within the brain stem consisting of a very complex network of small interconnecting neurones. Many nuclei are found within the region, most of which are small and unnamed. Of the recognized nuclei are the locus coeruleus and the raphe nuclei. As indicated in MCQ 176 reticular neurones lying largely in the medulla send descending fibres in the spinal cord (reticulospinal tract) to excite or inhibit motor neurones, thus forming part of the extrapyramidal system. Other areas are involved in the control of vegetative functions, including control of reflexes involved in maintenance respiration, heart rate and blood pressure. Other nuclei mainly located in the pons and midbrain send ascending axons to the thalamic cells, where they produce excitation or inhibition, and to the cortex; they also connect with the hypothalamus and limbic system. There are also afferent inputs from many sources. Somatovisceral afferents arrive by way of the spinoreticular tract and possibly the propriospinal pathways. Inputs also arrive from all the other cranial afferents. The function of the ascending reticular formation is not clearly understood. According to the reticular activity theory, signals from the reticular formation vary the activity of the thalamus and thus cortical neurone excitability. In this way arousal or excitation are produced. The reticular formation is also concerned with mediation of the affective–emotional aspects of sensory stimuli, particularly pain. TTTTT

MCQ 185

The electroecephalogram

1. during REM sleep resembles that of the wakeful subject.
2. shows bilateral asymmetry.
3. may show α waves if recorded in the relaxed awake state.
4. has waves of larger amplitude in the alert state than during sleep.
5. shows δ waves when the subject is exposed to a bright light.

An electroencephalogram (EEG) is a recording from the scalp and represents electrical activity arising from the numerous postsynaptic potentials occurring near the surface in response to cerebral activity as modified by inputs from subcortical structures; also activity arising from muscles and generator potentials in the eyes (retina). It is recorded by electrodes placed on the scalp. Conventionally eight to sixteen leads are used, arranged in a standard manner over specific cortical areas. In the normal awake and alert state the EEG shows irregular low-voltage waves of 30–80 μV. When the eyes are closed, but the subject is awake, an α rhythm is recorded, with a frequency of 8–12 Hz, although it is not seen in every subject. On falling asleep the EEG changes progressively from low-amplitude high-frequency waves of 10 Hz to θ waves (4–6 Hz) as sleep becomes deeper, and finally δ waves with a frequency of less than 4 Hz. Sleep is divided into slow-wave and rapid eye movement (REM) sleep. REM sleep is accompanied by signs of excitement and stress and the EEG tends to resemble that of the wakeful subject. If the subject is woken they report that they have been dreaming. The generation of the two types of sleep is not believed to be reticular, but under the control of the two brain stem nuclei: the locus coeruleus, whose cells secrete noradrenaline, and the raphe nuclei secreting 5-hydroxytryptamine. REM sleep is abolished by destruction of the locus coeruleus and both types by destruction of the raphe nuclei. Records of EEG may be of value clinically. The two main abnormalities are slow-wave or eleptiform. Large waves occur over the brain during an epileptic fit and abnormal patterns are frequently seen in these patients at other times. Abnormalities may also occur around tumours and hence aid in their localization. Asymmetry is a sign of illness. TFTFF

MCQ 186

The limbic system

1. includes the hypothalamus.
2. includes the amygdala.
3. includes the hippocampus.
4. is involved in the recognition of emotion.
5. is important in the control of posture.

The limbic system comprises a ring of cortical tissue on each side of the midline, encircling a number of structures which have a complex series of interactions. These structures include parts of the hypothalamus and thalamus, the septal nuclei, amygdaloid nuclei and hippocampus. The limbic system is associated with regulation of behaviour, being involved in analysis of olfactory signals, in feeding behaviour and control of several biological rhythms. It is also involved in sexual behaviour and emotional activity, fear and rage. Limbic system disease gives some indication of the function of the region, manifesting itself by effects on olfaction, memory and behaviour. The limbic system may be said to be the site at which emotional tone is generated. Thus specific hypothalamic areas, as identified using chronically implanted electrodes, are involved in both punishment and reward behaviour. For example, if a circuit is arranged so that a rat is able to stimulate its own brain via the electrodes the animal will repeatedly produce stimulation if it has a rewarding or pleasant effect. Stimulation of the amygdala produces aggressive behaviour while bilateral amygdalectomy results in an apathetic individual. The limbic system is also associated with the expression of emotion. Stimulation of the hypothalamus produces increased alertness, defensive postures, piloerection, dilated pupils and vocalization which has been described as sham rage. Stimulation of the amygdala can result in a number of individual movements associated with eating. The integrity of the limbic system is also essential for long-term memory. Damage to the hippocampus produces the greatest interference with memory and learning. Motor skills are not affected. TTTTF

MCQ 187

The cerebral cortex

1. receiving somatic information from the hands and face has a larger volume than that for the trunk and legs.
2. in the temporal region is associated with vision.
3. on the left side is usually dominant for speech in man.
4. has information exchanged between two hemispheres via the corpus callosum.
5. has association areas which are all those other than the primary sensory receiving zones.

The cerebral cortex is a region in the parietal lobe, associated with somatic sensations, which lies behind the central sulcus on a ridge known as the post-central gyrus. There is an orderly representation of the body along the sulcus, the area representing the leg lying medially and that representing the head lying laterally. Regions with the highest tactile discrimination have the largest volume of representation. The motor cortex in the frontal lobe occupies a region in front of the somatic sensory cortex, with each set of muscles supplied being represented in a particular region, those muscles having the most precise control (e.g. of the fingers and mouth) having the greatest area of representation. The temporal lobe of the cortex is associated with hearing, the occipital lobe with vision. The sensory association areas of the cortex exclude both the motor cortex and the primary sensory receiving zones, which serve for processing information. Examples of areas are the optical association area in the vicinity of the primary visual cortex, which is important for optical recognition, the parieto-occipital area (angular gyrus) responsible for the understanding of written words, and the acoustic association field (Wernicke's region), which is important for the understanding of speech. Broca's area (a triangular region of the inferior frontal gyrus) is concerned with the motor control of speech. It is only found in one hemisphere, the dominant hemisphere, which for 90 per cent of people is on the left. While cognitive functions are usually located in the left hemisphere, the right hemisphere predominates in such functions as spatial relationships in the external environment and musical forms. Information can be exchanged between the two hemispheres via the corpus callosum. Section of the corpus callosum, for example, results in a patient only being able to respond to verbal commands, with movement of muscles controlled by the dominant hemisphere for speech.
TFTTF

MCQ 188

Characteristic of short-term memory is

1. a relatively small capacity.
2. a duration of only days.
3. probable storage of material coded in words.
4. that it is probably represented in the nervous system by circulating nerve impulses.
5. that it is affected by electroconvulsive therapy and loss of consciousness resulting from head injury.

By definition memory is the acquisition of knowledge or skill learnt through instruction or experience. Learning involves storage of the information within memory as it is required. The human brain must be able to store information for several decades (long-term memory) or for a brief time lasting seconds to hours (short-term memory). The short-term memory has a small capacity and one theory is that it persists as a pattern of electrical circuits within the brain, as an insult which disrupts the functions of the neurones prevents memory of the events which occurred a short time before. Such insults include head injury, electroconvulsive therapy, coma and deep anaesthesia. The long-term mechanism must be different as the memories survive insults that disrupt neuronal activity.

Characteristic of the long-term memory is its large capacity and the organization which is semantic and relational. The mechanisms underlying long-term memory are not understood, but at least two have been proposed. One entails modification of the effectiveness of existing synapses and another the development of new synaptic contacts. Recently it has been shown that neurones can develop new synaptic terminals well into adult life. It has also been suggested as a result of studies using inhibitors that protein synthesis may be involved in long-term memory and attention has focused on macromolecules such as nucleic acids. TFTTT

Co-ordinated Functions of Body Systems

The hydrogen ion concentration

1. of a solution of pH 7 is 100 times greater than that of one of pH 4.0.
2. is doubled with a fall in pH of 0.3.
3. of a buffer solution remains constant if a strong acid is added to it.
4. is directly related to the hydroxyl ion concentration.
5. compatible with life is approximately 16–100 nmol l^{-1}.

By the Bronsted–Lowry definition of acids and bases an acid is a donor of hydrogen ions (H^+), while a base is a proton acceptor. Water may act as a proton donor through its partial dissociation into hydroxonium (H_3O^+) and hydroxyl (OH^-) ions. Only a fraction of water molecules, about one in 5.5×10^8 molecules, exist as ions. The ionization of water is usually written as

$$H_2O \rightleftharpoons H^+ + OH^-$$

and the product of hydrogen ion concentration and that of hydroxyl ions is constant

$$[H^+] \times [OH^-] = 10^{-14} = K$$

where K is the dissociation constant. In neutral solutions the numbers of hydrogen and hydroxyl ions are equal, both being 10^{-7}. Thus in an acid solution of 10^{-4} M the hydroxyl ion concentration will be 10^{-10} ($10^{-4} \times 10^{-10} = 10^{-14}$). The ion product of water is the basis of the pH scale. One takes the exponent of the hydrogen ion concentration and makes it positive; thus pH is the negative log to the base 10 of the hydrogen ion concentration. The hydrogen ion concentration of a solution of pH 7.0 is 10^{-7} M and that of a solution of pH 4 is 1×10^{-4} M. A solution of 0.005 M hydrogen ions or 5×10^{-3} will have a pH of 2.3, i.e. −3.0 + 0.7 (or the log of 5). A solution of double the hydrogen ion concentration, namely 0.01 M, is of pH 2.0. Thus because a log scale is used the pH decreases by 0.3 for a doubling of the hydrogen ion concentration. In physiological terms the normal pH is 7.4. Below 7.35 acidosis is said to occur, even though the pH is neutral. The range of pH compatible with life is 7.0–7.8, corresponding to a hydrogen ion concentration of 100–16 mmol l^{-1}. A buffer solution is one which tends to diminish, not prevent, the change in pH which would be produced by addition of exogenous acid or alkali. It consists of a weak acid and the salt of the acid. The weak acid acts as a reservoir of hydrogen ions which can neutralize a base. The base provided by the salt of the acid can react with hydrogen ions to form the weak acid, thus neutralizing added acid. FTFFT

MCQ 190

In the maintenance of hydrogen ion concentrations in the body

1. intracellular protein plays a role.
2. haemoglobin is a buffer involved.
3. phosphate is quantitatively more important than plasma protein.
4. if carbon dioxide rises as a result of depressed respiration, then bicarbonate reabsorption is increased.
5. bicarbonate is an important buffer as the pK is close to the pH of the extracellular fluid.

The pH of the blood and body fluids is constantly threatened by, for example, the metabolic production of carbon dioxide and other acidic wastes. The first defence against shifts of blood pH comprises blood and tissue buffers which are immediately effective. Second there is respiratory compensation, acting in a period of minutes, and finally renal compensation, acting over a period of hours. The pH of a buffer solution is given by the Henderson–Hasselbalch equation:

$$\mathrm{pH} = \mathrm{pK} + \log \frac{[\mathrm{salt}]}{[\mathrm{acid}]}$$

When pH equals pK the concentrations of salt and acid are equal and the buffer acts more efficiently. Some regard the most important buffer system of the body as being the protein system, because proteins occur in the cytoplasm and extracellular fluids, including the plasma proteins. Haemoglobin is also an effective buffer, partly because of the histidine groups. Proteins in general are effective buffers as they contain numerous ionic groups such as amino and carboxyl groups which interact reversibly with hydrogen ions. Phosphate buffers consist of sodium and potassium salts of phosphoric acid that occur in blood, body fluids and urine. Hydrogen ions can replace sodium or potassium from the monohydrogen salt to from dihydrogen phosphate. This is important in buffering of urine pH, but only of limited importance in buffering of blood and body fluids because of the relatively few hydrogen ions which can be donated or accepted. The carbonic acid–bicarbonate buffer system is also very important. Indeed Henderson regarded it as the most important. It works closely with the haemoglobin–oxyhaemoglobin buffers to keep the arterial and venous blood from varying greatly in pH. The ratio of bicarbonate to carbonic acid is 20 : 1 and the pK is 6.1, so that it would not appear to be an effective system. However, this is an example of a volatile buffer system in which one component (H_2CO_3) is in equilibrium with a gas, CO_2. The $P\mathrm{CO}_2$ is maintained constant via respiration even with a large amount of acid and base added. Equally, bicarbonate can be controlled by the kidney. TTFTF

MCQ 191

A respiratory acidosis

1. is produced by an overdose of salicylate.
2. is seen in a high fever.
3. is produced by an overdose of barbiturates.
4. is associated with an increase in plasma bicarbonate.
5. is associated with primary aldosteronism.

Primary disturbances in respiratory function which alter arterial $P\text{CO}_2$ cause disturbances in acid–base balance. Impairment of alveolar ventilation results in increased $P\text{CO}_2$ and hence arterial pH rises. Respiratory acidosis may be produced by depression of the medullary respiratory centres, as seen with an overdose of barbiturates, tranquillizers or morphine, by reduced airflow as seen for example in severe asthma, emphysema or pneumonia, or by restriction of normal ventilation as in gross obesity or paralysis. The pH is kept relatively constant by buffering and increased bicarbonate, which helps to keep the ratio of bicarbonate to carbonic acid (which represents dissolved CO_2) constant. The kidney also contributes by secreting in exchange for sodium ions hydrogen ions, which are formed together with bicarbonate from carbonic acid. Within the tubule hydrogen ions are eliminated by combination with bicarbonate ions, phosphate salts or ammonium ions. The ammonia is formed from glutamine under the influence of glutaminases in the renal tubule (see MCQ 90). Respiratory alkalosis is brought about by stimulation of the respiration as in hysterical hyperventilation or in fever or by stimulation of chemoreceptors, of pulmonary J receptors or on passive mechanical ventilation. Such patients may have a $P\text{CO}_2$ of less than 30 mmHg (4.0 kPa), which results in decreased cerebral blood flow and lightheadedness. The increased pH of the body fluids results in a decreased concentration of ionized calcium and hence altered excitability of muscle. There is also a fall in plasma bicarbonate which is enhanced by the action of the kidney. Primary aldosteronism with excess secretion of the mineralocorticoid and pituitary deficiency results in metabolic alkalosis resulting from a shift of extracellular fluid hydrogen ions into the cells and excretion of an acid urine. FFTTF

MCQ 192

In a patient with complete pyloric obstruction resulting in severe vomiting there may be

1. an increase in arterial pH.
2. an increase in plasma bicarbonate concentrations.
3. stimulation of respiration.
4. the production of a relatively acid urine.
5. increased plasma chloride concentrations.

Loss of gastric acid, as in prolonged vomiting, results in a metabolic alkalosis, i.e. one of non-respiratory origin. This may also be produced by over-zealous self-medication with bicarbonate or an increased secretion of hydrogen ions by the renal tubule as a result of diuretic therapy. Protracted vomiting is the most important cause of metabolic alkalosis, potassium deficiency being the second. As well as an increased pH, a characteristic feature is an increased bicarbonate concentration in the extracellular fluid. A normal $HCO_3^-:H_2CO_3$ ratio can be partly re-established by a decrease in ventilation, so that in contrast to respiratory alkalosis there is an initial increase in the arterial $P\text{CO}_2$. In the metabolic alkalosis of non-renal origin the kidneys also respond to an increase in plasma bicarbonate by increased bicarbonate excretion. Initially, however, it is possible that the loss of sodium from the body results in renal sodium retention which promotes the excretion of hydrogen ions in the urine. Thus electrolyte balance is maintained in the face of an adverse acid–base balance. Metabolic acidosis in contrast is characterized by a fall in pH and a large fall in plasma bicarbonate. In only some instances is the cause truly metabolic. It may be produced by an increase in the concentration of strong, non-volatile acid through excessive production or inadequate secretion. Alternatively there may be loss of bicarbonate, either by loss of intestinal contents as in diarrhoea or by inadequate renal absorption. Excess acid production occurs through production of lactic acid and, in conditions of increased anaerobic glycolysis, following impaired tissue oxygenation or through incomplete oxidation of fatty acids resulting in the ketoacidosis of diabetes mellitus. It can also be caused by ingestion of acid or from agents metabolized to acid products. As the excess acid is buffered by the bicarbonate, carbonic acid is formed, which dissociates into water and carbon dioxide. The latter is blown off as respiration is stimulated. Further stimulation of respiration may be brought about by stimulation of the peripheral chemoreceptors. Stimulation of respiration may help restore pH, but does not rid the body of excess hydrogen ions. These are excreted through the kidney. TTFTF

MCQ 193

Five minutes after a 70-kg individual loses one litre of blood there would be

1. a reduced diastolic pressure.
2. a decreased total peripheral resistance.
3. splanchnic vasodilatation.
4. an elevated glomerular filtration rate.
5. a decreased venous tone.

The circulating blood volume is about 5 l and a haemorrhage of up to one litre can be compensated for, depending on the rate of bleeding. Haemorrhage produces a decreased venous tone, a fall in atrial pressure, and a reduction in end diastolic volume, resulting in a reduced stroke volume and hence a fall in blood pressure. This in turn leads to stimulation of the baroreceptors and atrial receptors. The chemoreceptors are also stimulated. These events result in increased activity in the centres concerned with vasomotor and cardiac control, with a consequent fall in the capacity of the circulation and an increase in the heart rate and cardiac contractility. The release of catecholamines augments these reflex changes, and the release of vasopressin and the production of angiotensin II enhance the vasoconstriction. The tissues most affected are those such as the skin, kidney and gut. The increased sympathetic efferent activity to skin causes marked pallor and inappropriate sweating, so that the patient feels cold and clammy. Haemorrhage causes vasoconstriction of both afferent and efferent renal arterioles, which produces a decreased glomerular filtration rate and oliguria. Venous constriction is an important response in haemorrhage, which serves to keep the volume in the systemic arteries near normal at the expense of the venous blood volume. For this reason the blood in the veins is regarded as part of a dynamic reservoir for the arteries. In most healthy subjects five minutes after a moderate haemorrhage these venous and arteriolar changes combine to keep the mean systemic pressure near normal. There may be reduced systolic pressure but not diastolic pressure. FFFFF

MCQ 194

Following the loss of one litre of blood in a 70-kg individual there is

1. a decreased concentration of plasma proteins.
2. a decreased capillary hydrostatic pressure.
3. sodium retention.
4. a decreased secretion of vasopressin.
5. an increased concentration in plasma of angiotensin II.

Following haemorrhage, not only are there changes which reduce the capacity of the circulation, but also some which increase the filling. Fluid moves into the cardiovascular system from the interstitium. The fall in blood pressure and the accompanying arteriolar vasoconstriction results in a fall in the capillary hydrostatic pressure and hence the filtering pressure and the fluid lost from the cardiovascular system is reduced. This leads to a dilution of the plasma proteins, a change which would, however, reduce the fluid uptake. This capillary mechanism only results in redistribution of the extracellular fluid. Fluid loss must also be reduced. The fall in blood pressure contributes to a reduced glomerular filtration rate and oliguria. The decreased blood volume also produces an elevation of vasopressin and hence enhanced fluid retention. There is also increased renin release from the renal juxtaglomerular apparatus and hence of angiotensin II formation, resulting in increased output of aldosterone from the adrenal cortex. This leads to a positive sodium balance and hence increased extracellular fluid volume. Inhibition of release of atrial natriuretic peptide may also reduce sodium output and hence contribute to the positive sodium balance. A further mechanism in restoring plasma volume is thirst and increased fluid intake. Hence administration of fluid is of value, but not alcohol as it produces peripheral vasodilatation and reduces the blood pressure. The same is true of covering the patient with too many blankets. TTTFT

MCQ 195

During the long-term recovery from haemorrhage

1. plasma volume is restored in under 12 h.
2. the intestinal absorption of iron is increased.
3. there is increased synthesis of plasma protein in the liver.
4. red blood cells are replaced in 4–8 weeks.
5. there is an increase in the number of circulating reticulocytes.

After a haemorrhage there is a need for replacement of circulating plasma, namely salt and water and plasma proteins, and of the cellular contents, especially red cells. Through a shift of fluid between compartments, reduced salt and water output and enhanced fluid intake, plasma volume is restored within a period of 12–72 h. After the fluid has been replaced, the reduced oncotic pressure would prevent further absorption of water and hence full compensatory expansion. However, some plasma protein is replaced, first, by interstitial protein carried by the lymphatics to the vascular compartment as a result of increased lymph flow and, second, there is increased synthesis of plasma protein in the liver brought about as a result of the dilution of the plasma with interstitial fluid. This is accomplished within 3–4 days. The red blood cells are restored within 4–8 weeks. Reduction of the blood flow to the kidney stimulates the release of erythropoietin, a hormone (as indicated in MCQ 13) which increases the rate of maturation and release of red blood cells into the bloodstream. A precursor to the erythrocytes is the reticulocyte, which increases in concentration in the blood during times of increases in red cell production. The increased production of erythrocytes results in increased utilization of iron for haemoglobin production. The removal of iron from the plasma leads to an increase in the absorption of iron from the gastrointestinal tract. In cases of severe haemorrhage rapid transfusion of blood is required. Saline alone is of relatively little value because it rapidly distributes through the entire extracellular compartment. Plasma expanders which remain in the cardiovascular space are of some use in maintaining the circulation. However, serum albumin tends to dehydrate tissue by drawing out fluid. FTTTT

MCQ 196

During exercise

1. there is an increase in arterial $P\text{CO}_2$.
2. respiration is increased via stimulation of peripheral chemoreceptors.
3. cardiac output can increase ten times.
4. diastolic blood pressure increases.
5. diastole shortens more than systole.

During exercise the rate of metabolism of muscle cells may increase up to 20 times, with a resultant fall in $P\text{O}_2$ and pH and an associated increase in temperature. These changes and possibly also accumulation of other products of metabolism such as adenine and potassium ions leads to vasodilatation, with an increase in the blood flow of up to 18 times. To sustain such a fall in resistance in muscle vessels extensive cardiovascular changes are required and these are increased cardiac output and redistribution of blood flow. The cardiac output may increase from 5 to 35 l min^{-1} in the trained athlete. This is a result of increased stroke volume and heart rate, which may go up to 200 beats min^{-1}, as a result largely of shortening of diastole. The changes result from increased activity in the sympathetic outflow. This increased activity also leads to vasoconstriction in a number of vascular beds, including those of the renal and splanchnic regions. There is obviously vasodilatation of the coronary vessels associated with the increased cardiac work and in the skin in order to dissipate the heat produced as a result of exercise. The major heat loss, however, is via increased secretion or evaporation of sweat. The cardiovascular changes tend to leave the diastolic pressure unchanged. There is a tendency for the systemic pressure to increase, so that the mean arterial blood pressure is unchanged or increases slightly. There is, however, an increase in pulse pressure. Pulmonary ventilation increases as the muscle requirement for oxygen is increased. How this is achieved is not clear as no changes in $P\text{CO}_2$, $P\text{O}_2$ or pH are observed at the time of the increase. It has therefore been suggested that there is a drive from the higher centres, the cerebral cortex irradiating information to the medulla. It is also possible that this is a learned response. FFFFT

MCQ 197

During severe exercise

1. there is an increase in the circulating lactic acid.
2. there is a decrease in venous PO_2.
3. glycogen stores are depleted.
4. blood glucose concentrations immediately fall rapidly.
5. there is reduced gluconeogenesis.

The severity of exercise can be expressed in terms of the energy cost or oxygen consumption or in terms of the rate at which work is performed, both being expressed in terms of kJ min^{-1}. The energy requirements for a 70-kg adult asleep are 5.07 kJ min^{-1}, up to around 272–278 kJ min^{-1} during horizontal running at 8 mph. The efficiency with which a human works is given by

$$\text{efficiency (\%)} = \frac{\text{rate of work done}}{(\text{oxygen used} \times \text{calorific value of oxygen})}$$

The oxygen used is that used in exercise minus that at rest. The efficiency depends on a number of factors, including speed, load, fatigue and training. The average efficiency is 20–25 per cent, similar to that of the internal combustion engine, but greater than that of the steam engine, at 10 per cent. The oxygen uptake at the beginning of exercise does not increase sufficiently to allow the energy requirements to be supplied by aerobic metabolism, so that there is a decrease in the PO_2 of venous blood and myoglobin oxygen stores are depleted. The stores of ATP and creatine phosphate are also depleted. If the exercise is severe or near maximal for the subject ATP is also produced by the breakdown of muscle glycogen and glucose to lactic acid. Blood glucose is maintained initially, implying a greater rate of production than that of utilization. Subsequently it may decrease. On stopping work the oxygen consumption does not fall to resting levels immediately. The oxygen utilization over and above the resting level during the period is theoretically equal to the oxygen debt accumulated at the start of exercise. The oxygen debt is used to replenish ATP and creatine phosphate stores, myoglobin oxygen stores and to remove the lactate produced. TTTFF

MCQ 198

On ascent to high altitude there is

1. increased arterial $P\text{CO}_2$.
2. stimulation of respiration.
3. enhanced renal reabsorption of bicarbonate.
4. vasodilation of cerebral vessels.
5. elevated pulmonary vascular resistances.

On ascent to high altitude there is a reduction in barometric pressure and hence oxygen tension. Below 3000–4000 m (10000–12000 ft) the arterial $P\text{O}_2$ is still on the plateau of the oxygen dissociation curve and there is little immediate change in ventilation. When the arterial pressure falls below 60 mmHg (8 kPa) respiration is stimulated through the mediation of peripheral chemoreceptors. During exercise, however, the minute volume increases markedly. The hypocapnia resulting from the hyperventilation in turn inhibits ventilation and is associated with respiratory alkalosis, which is partly corrected by the lactic acid and pyruvic acidaemia seen. Later renal compensation becomes effective. The alkalosis is also associated with a leftward shift of the oxygen dissociation curve, which facilitates binding in the lung but makes it more difficult for oxygen to be released in the tissues. Hyperventilation may also lead to dehydration because of the water loss via the lungs. There are also cardiovascular changes, with an increase in heart rate and cardiac output, although arterial blood pressure is altered little at rest. In the systemic circulation hypoxia is a major vasodilator and hypocapnia has a relatively slight effect. However, in the lungs hypoxia is a potent vasoconstrictor, and in the brain hypocapnia results in vasoconstriction. If lowlanders ascend too rapidly to altitude, they may show exaggerated responses, with pulmonary hypertension leading to pulmonary oedema. Increasing hypoxia impairs psychomotor performance progressively. At lower altitudes (3000 m) the only objective functions known to be affected concern the learning and performance of novel tasks. Above 4000 m there are clear deteriorations in sensory acuity, vigilance, judgement, speed of response and manual dexterity. The combination of dizziness, headache, vomiting, nausea, dyspnoea (difficulty in breathing) and inco-ordination constitutes a condition described as acute mountain sickness. FTFFT

MCQ 199

On acclimatization, the work capacity at altitude

1. is comparable to that of subjects normally living at altitude.
2. depends on an increased packed cell volume.
3. results in part from a high diffusing capacity in the lung.
4. requires an increased cardiac output.
5. is associated with a high $P\text{O}_2$ gradient across the lung.

Theoretically hypoxia may be overcome by turning to aerobic processes, by increasing the affinity of existing aerobic processes or by easing the access of oxygen to the tissues. Research has concentrated on the latter and a number of mechanisms have been identified which contribute to the acclimatization which allows people to live successfully up to an altitude of 6000 m (18000 ft). Hyperventilation is maintained, but is not entirely due to the hypoxic drive as it is maintained for a time after rapid descent. It appears that the reduced hydrogen ion concentration, plasma and cerebral interstitial fluid produced by hyperventilation lead to increased transport of bicarbonate out of the cerebral fluid and interstitial fluid so that the $P\text{CO}_2$:HCO_3 ratio and the hydrogen ion concentration are brought back to normal. As there is relatively little protein to buffer hydrogen ions in the cerebrospinal fluid, a reduction in ventilation as seen when the individual is exposed to normal oxygen tensions results in an increased hydrogen ion concentration, which stimulates respiration. The systemic alkalosis developing on ascent to high altitude is corrected over a period of two to three weeks by excretion of bicarbonate by the kidneys. Oxygen transport to the tissues is increased as a result of increased red cell and haemoglobin production, so that the packed cell volume may be 60 per cent or more. The resultant increase in viscosity increases the demands on the heart. Further demands are made by the hypoxic vasoconstriction in the vascular bed of the lungs, where there is also increased muscularity of the arterial walls. The increased load results in right ventricular hypertrophy. Cardiac output and heart rate return to normal on acclimatization, but the maximum workload is decreased. Increased vascularity of tissues results in a decreased supply of oxygen to the tissues, an effect enhanced by the elevation of 2,3-diphosphoglycerate, which shifts the dissociation curve to the right so that oxygen is given off more readily in the tissues. FTTTF

MCQ 200

The maintenance of body temperature

1. means that it is kept constant over 24 h.
2. involves medullary thermoreceptors.
3. involves parasympathetic fibres to the skin vessels.
4. may involve sweating.
5. involves thermoreceptors in the skin.

The temperature of the superficial shell of the body varies considerably. The temperature of different tissues within the body also varies, but the core temperature remains within narrow limits. It does not, however, remain constant over the day; in most subjects the temperature on waking is lower than the temperature on retiring. The relatively constant temperature is maintained by a balance between the production of heat through metabolism and loss on radiation, conduction, convection and evaporation. Temperature receptors are present both in the periphery and in the hypothalamic area. The former signal changes in the skin and the latter changes in the temperature of the local blood supply. Both peripheral and hypothalamic sensors are axonally connected to a neuronal network in the anterior hypothalamus and preoptic region. Some six pathways have been identified through which the hypothalamus can induce thermoregulatory responses. First, adrenergic non-medullated fibres of the sympathetic nervous system bring about vasoconstriction and there is probably some inhibition of sympathetic vasodilator nerves. The pilomotor reaction is also brought about by adrenergic sympathetic nerves. Second, the response to heating involves active cutaneous vasodilatation via the sympathetic system and inhibition of vasoconstrictor tone. Third, shivering is induced by the thermal centre in the posterior hypothalamus, the normal motor nerve supply being involved. Fourth, the sympathetic supply to the adrenal medulla is stimulated so adrenaline is released, which increases heat production and cardiac output, enhancing skin vasoconstriction and promoting dilatation of muscle blood vessels. Fifth is non-shivering thermogenesis, which involves the sympathetic system and has been claimed to occur in man. Finally, sweating is activated through cholinergic sympathetic nerves. With prolonged exposure and acclimatization to heat and cold the effector mechanisms which come into play involve endocrine systems such as the pituitary–thyroid, pituitary–adrenal axes and the neurohypophysis.
FFFTT

MCQ 201

In the development of a fever

1. pyrogens are the only cause.
2. the hypothalamic 'thermostat' is reset.
3. the patient feels hot.
4. radiant heat loss is reduced.
5. very little water is lost through sweating.

Body temperature varies over the day, being highest during the evening hours. If it exceeds the normal range, hyperpyrexia or fever is said to occur. Common causes are infection or primary neurological disorders. Substances which produce fevers are called pyrogens, bacterial pyrogens being endotoxins from gram-negative bacilli and endogenous pyrogens being derived from leucocytes, possibly the membranes. The exact mechanism of action of these agents is unknown, but they are believed to elevate the temperature reference point. Since aspirin—an agent which inhibits prostaglandin synthetase—lowers temperature, prostaglandins may be involved. The fever progresses through four stages. Initially, during the prodromal phase the patient feels generally unwell and there is piloerection. This is because on resetting the hypothalamic 'thermostat' the patient responds as if placed in a cold environment. In the second stage—the chill—the patient feels cold, despite the rising body temperature, and shivering occurs to generate additional heat. This continues until the body temperature reaches the new set point. In the third stage, the flush, the patient feels warm and there is vasodilatation in the skin. Regulation at the new point occurs until the so-called crisis. In the final stage, defervescence, when the set point has been shifted back to the normal level, heat loss occurs via sweating. The increased rate of metabolism may speed the healing of injured tissues. It is also possible that the fever reduces the resistance of the infecting micro-organism to the host's defence mechanisms. However, prolonged fever can injure the cells of the body, especially the nervous cells. Hyperthermia may also accompany heat stroke, in which the body cannot be cooled by evaporation, as in hot humid weather. Extensive vasodilatation occurs, which may lead to hypovolaemic shock. This contrasts with heat exhaustion which results from loss of fluid and electrolytes on excessive sweating with physical exercise. FTFTT

MCQ 202

Under high-pressure conditions during diving

1. the arterial partial pressure of nitrogen is greater than normal.
2. arterial $P\text{CO}_2$ is much greater than normal.
3. nitrogen may produce narcosis.
4. the resting lung volume is reduced.
5. pure oxygen is toxic.

Divers working at depth are faced with many problems, a major one being the increased ambient pressure; for every 10 m (30 ft) the pressure increases by one atmosphere. There is no change in the lung volume as the air in the lungs is at the same pressure as the surroundings. The only condition when this is not true would be a rapid descent performed with inadequate quantities of gas. Because of the hyperbaric conditions the arterial tensions of inspired gases are higher than normal, whereas the arterial $P\text{CO}_2$ is unaffected, as it is controlled by respiration. If nitrogen is breathed at high pressures for a period a large quantity of nitrogen becomes dissolved in the body fluids and tissues, especially fatty tissues. Nitrogen is not metabolized in the body, so that as the diver returns to the surface the nitrogen comes out of solution. The bubbles formed cause considerable damage (decompression sickness). They will form emboli in the pulmonary, myocardial and cerebral circulation or may be lodged in the joints, leading to dyspnoea, paralysis and severe pain in the joints. The phenomenon of supersaturation allows nitrogen to remain dissolved if the pressure is not three times that of the body. This means that a diver can ascend from 20 m (66 ft) with no harmful effects. Nitrogen at high pressures can also produce narcosis, dissolving in the membranes and other lipid structures of the neurones and thus reducing their excitability. Nitrogen is therefore generally replaced by helium as it has a smaller narcotic effect. Oxygen at high pressure can also be detrimental to the central nervous system and may cause convulsions. Carbon dioxide at high pressure can depress respiration and cause acidosis, but with well-designed equipment this should not occur. Immersion in water also makes it harder to maintain thermal equilibrium, because water conducts heat about 30 times more effectively than air. Perception is also affected. Apart from darkness, the indices of refraction of light travelling from water, through glass, to the air space in front of the eye makes the image appear at 75 per cent of the actual distance. Water-borne sound impinges on the head and hearing is assumed to be via bone conduction. Binaural clues are also absent. TFTFT

Index